Quantum Science and Technology

The book series Quantum Science and Technology is dedicated to one of today's most active and rapidly expanding fields of research and development. In particular, the series will be a showcase for the growing number of experimental implementations and practical applications of quantum systems. These will include, but are not restricted to: quantum information processing, quantum computing, and quantum simulation; quantum communication and quantum cryptography; entanglement and other quantum resources; quantum interfaces and hybrid quantum systems; quantum memories and quantum repeaters; measurement-based quantum control and quantum feedback; quantum nanomechanics, quantum optomechanics and quantum transducers; quantum sensing and quantum metrology; as well as quantum effects in biology. Last but not least, the series will include books on the theoretical and mathematical questions relevant to designing and understanding these systems and devices, as well as foundational issues concerning the quantum phenomena themselves. Written and edited by leading experts, the treatments will be designed for graduate students and other researchers already working in, or intending to enter the field of quantum science and technology.

More information about this series at http://www.springer.com/series/10039

Federico Grasselli

Quantum Cryptography

From Key Distribution to Conference Key Agreement

 Springer

Federico Grasselli 🆔
Heinrich Heine University Düsseldorf
Düsseldorf, Germany

ISSN 2364-9054   ISSN 2364-9062   (electronic)
Quantum Science and Technology
ISBN 978-3-030-64359-1   ISBN 978-3-030-64360-7   (eBook)
https://doi.org/10.1007/978-3-030-64360-7

This Springer imprint is published by the registered company Springer Nature Switzerland AG
The registered company address is: Gewerbestrasse 11, 6330 Cham, Switzerland

To my parents, Valeria and Sandro

Preface

Stimulated by concerns about the secrecy of our data and their secure transmission, there is worldwide active research on quantum cryptography and particularly on Quantum Key Distribution (QKD).

QKD allows two users to communicate with a higher level of security thanks to the laws of quantum physics, making it a candidate to replace current cryptographic schemes. This attracted the attention of governments, institutions and companies to invest in quantum cryptographic technologies, as confirmed by regular theoretical breakthroughs and record-breaking experiments.

This book intends to introduce the reader to QKD and its recent multiparty generalization, called (quantum) Conference Key Agreement (CKA).

While providing a comprehensive view of the state of the art of the field, the book also offers a didactic guidance towards the latest developments, allowing non-expert readers to eventually participate in the on-going research. This is eased by explaining new ideas in an intuitive way and by going into the details of fundamental concepts.

I believe that the growing attention being received by QKD, combined with the novel introduction to CKA, makes this book a timely publication.

I wish the reader an enjoyable read and I hope it can inspire further progress on the flourishing field of quantum cryptography.

Düsseldorf, Germany   Federico Grasselli
November 2020

Acknowledgements

Without the knowledge and expertise developed during my doctoral studies, as well as the advancements we contributed to, this book would probably not have been possible. For this, I must thank my Ph.D. supervisor Dagmar Bruß and my co-supervisor Hermann Kampermann for giving me the opportunity to join them at the "Institute for Theoretical Physics III" in Heinrich-Heine-Universität Düsseldorf. My research greatly benefited from the guidance provided by Dagmar and Hermann, from their constructive criticisms and their passionate attitude towards science. They will certainly represent an inspiration for me throughout my career.

I also would like to thank my other colleagues in Düsseldorf. In particular, I acknowledge the fruitful collaborations with Gláucia Murta and the help of Carlo Liorni, Giulio Gianfelici, Lucas Tendick, Sarnava Datta and Thomas Wagner who proofread some chapters of this book.

I am grateful for the opportunities to collaborate with other great scientists, who enriched and widened my view on many research topics. Among these, I am particularly thankful to Marcos Curty and Álvaro Navarrete in Vigo and Massimiliano Proietti, Joseph Ho and Alessandro Fedrizzi in Edinburgh.

Moving from Italy to Germany and starting a new social life was made much easier thanks to the amazing group of international friends I had the fortune to meet in these years. A special thanks goes to Ayesha Din for her love, unconditional support and for being my biggest fan.

The greatest merit for shaping the person I am and enabling me to achieve such goals goes to my parents. The value of the education I received from them becomes clearer—as for many of us—when older. Here I acknowledge all of that value and express my deepest gratitude.

Contents

Acronyms

AEP	Asymptotic Equipartition Property: links the smooth min- and max-entropy of an i.i.d. state to the von Neumann entropy.
BB84	Bennett-Brassard 1984 (protocol): arguably the earliest and most popular QKD protocol, devised by Bennett and Brassard in 1984.
BS	Beam Splitter: optical device that splits incoming light into a transmitted and a reflected beam.
CHSH	Clauser-Horne-Shimony-Holt (inequality): a bipartite Bell inequality.
CKA	(Quantum) Conference Key Agreement: the task of establishing a secret key among several parties, the conference key, with quantum resources.
DI	Device-independent (protocol): the security of the intended protocol holds independently of the functioning of the employed quantum devices (sources, measurement apparatuses).
DICKA	Device-independent (Quantum) Conference Key Agreement.
DIQKD	Device-independent Quantum Key Distribution.
DIRG	Device-independent Randomness Generation.
EAT	Entropy Accumulation Theorem: theoretical result that quantifies the entropy accumulated in sequential rounds of a DI protocol. It generalizes the AEP to non-i.i.d. scenarios.
EC	Error Correction: the classical task of correcting the errors affecting a bitstring that is supposed to coincide with another bitstring.
GHZ	Greenberger–Horne–Zeilinger (state): a GME state that is commonly used in quantum cryptography.
GLLP	Gottesman-Lo-Lütkenhaus-Preskill (proof): a security proof for QKD protocols implemented with WCPs.
GME	Genuine Multipartite Entangled (state): a multiparty quantum state that cannot be written as a convex combination of biseparable states.
HOM	Hong-Ou-Mandel (effect): optical interference effect experienced by two indistinguishable photons.

HWP	Half-wave Plate: optical device that alters the polarization of light.
i.i.d.	independent and identically distributed.
KG	Key Generation (round): the intended round of a QKD protocol is devoted to the generation of a key bit.
LHV	Local Hidden Variable (model): any physical model whose probability distributions assume Bell locality and no-signaling constraints at the microstate level.
MABK	Mermin-Ardehali-Belinskii-Klyshko (inequality): a multipartite generalization of the CHSH inequality.
MDI-QKD	Measurement-device-independent QKD: a QKD paradigm where the security of the protocol does not rely on the trustworthiness of the quantum measurement devices.
PA	Privacy Amplification: the classical task of enhancing the privacy (secrecy) of a bitstring.
PBS	Polarizing Beam Splitter: optical device that splits incoming light into two beams with orthogonal polarizations.
PE	Parameter Estimation: a classical post-processing step of QKD where the honest parties estimate the noise affecting the quantum channel.
PLOB	Pirandola-Laurenza-Ottaviani-Banchi (bound): an upper bound on the private capacity of a pure-loss bosonic channel.
PNS	Photon Number Splitting (attack): an attack on a QKD system performed by an eavesdropper who wants to learn the secret key.
POVM	Positive Operator-Valued Measure: the most general measure in quantum mechanics.
PST	Post-selection Technique: a theoretical result linking the security of a QKD protocol under collective attacks to its security under coherent attacks.
QBER	Quantum Bit Error Rate: the error rate affecting two bitstrings obtained as a result of the same quantum measurement performed on two different sets of quantum systems (e.g. Alice's systems and Bob's systems).
QKD	Quantum Key Distribution: the task of establishing a secret key among two parties with quantum resources.
QWP	Quarter-wave Plate: optical device that alters the polarization of light.
SPD	Single-photon Detector: device capable of detecting the arrival of just one photon.
SPDC	Spontaneous Parametric Down Conversion: optical process that converts a photon of high energy entering a non-linear crystal into two photons of lower energy.
TF-QKD	Twin-field QKD: a QKD approach based on single-photon interference between optical fields with the same random phase (twin fields).
VOA	Variable Optical Attenuator: optical device that modulates the intensity of light.

WCP   Weak Coherent Pulse: optical pulse typically generated by highly
               attenuated lasers and largely used for QKD implementations.
WDM   Wavelength Division Multiplexing: a technique commonly used in
               telecommunication infrastructures to transmit several optical signals
               over the same optical fibre.

Chapter 1
Introduction

God had meant photons to travel rather than to stay put! This was the insight that made us think of using a quantum channel to transmit confidential information. Gilles Brassard

Quantum cryptography exploits distinctive quantum properties of nature in order to perform a given cryptographic task. Most quantum cryptographic protocols are—at least in principle—information-theoretically secure, which is a very strong notion of security as it is deduced purely from information theory.

Early ideas to use quantum properties for security purposes date back to the 70's [1, 2], when Wiesner aimed to create unfalsifiable bank notes [3]. These ideas seemed however very unpractical as they required to store a single polarized photon for days without losses (at the time, photon polarization was the only conceived carrier of quantum information).

The breakthrough occurred in 1983, when Bennett and Brassard realized that photons are best used to transmit quantum information rather than to store it. In particular, they could be used to transmit a random secret key from a sender to a receiver, who can then use the key to encrypt and decrypt sensitive messages. Shortly after, Bennett and Brassard published the first quantum key distribution (QKD) protocol in 1984 [4], hence named BB84 protocol. A QKD protocol enables two parties to generate a shared secret key via an insecure quantum channel and an authenticated public classical channel. Since then, many new protocols have been proposed [5–7] and implemented [8], allowing QKD to become the flagship of quantum cryptography and one of the major applications of quantum information science.

Furthermore, pushed by increasing concerns on data security and by the prospect of commercialization, the research on quantum cryptography has spread beyond the walls of academia and attracted the attention of several companies, private institutions and governments [9, 10]. In fact, a growing number of companies and start-ups worldwide are offering quantum cryptographic solutions.

© The Author(s), under exclusive license to Springer Nature Switzerland AG 2021
F. Grasselli, *Quantum Cryptography*, Quantum Science and Technology,
https://doi.org/10.1007/978-3-030-64360-7_1

1

In the long term scientists envision the creation of large-scale quantum networks where, thanks to quantum entanglement, QKD-enabled secure communication will be possible among any subset of users in the network. With a broader perspective, such networks could be linked together in a quantum internet [11, 12] that would serve much more scopes than just secure communication, e.g.. secure access to remote quantum computers [13, 14].

On the one hand, this book aspires to be a beginner's guide to QKD and other related topics. Nonetheless, it aims at carrying the reader from the fundamentals of the subject up to the latest research results. The book is characterized by a pedagogical approach, enriched by intuitive explanations and detailed calculations, sometimes hard to find in published papers. The ultimate goal is to prepare the reader to eventually take part to the research efforts on this flourishing field.

1.1 Background and Motivation

Quantum cryptography beautifully combines ideas and contributions coming from various fields of study, ranging from quantum information and quantum communication, to computer science and classical cryptography. The interplay between these diverse disciplines leads to theoretical advancements that can be of broad interest and applicable to other research fields.

Nevertheless, because of the significant commercial appeal of quantum cryptography and in particular of QKD, the on-going research is also guided by more practical purposes. For instance, combined theoretical and experimental efforts are constantly devoted to: stretc.hing the maximum distance at which QKD can be performed, increasing the key-generation rates, simplifying the experimental setups, and so on.

To this aim, the book addresses a novel QKD protocol that has recently received a lot of attention from the scientific community and is considered to be the new benchmark for long-distance QKD in fibre. The protocol, named twin-field (TF) QKD [15, 16], allows two parties to establish a secret key over long distances with single-photon interferometric measurements occurring in an intermediate relay. In this context, we analyse in detail realistic implementations of TF-QKD by employing recent theoretical results and simulations [17, 18].

With bipartite QKD links becoming the norm in several research centres and on-field installations all over the world, the next big step would be to interconnect such isolated links into quantum networks in order to perform more sophisticated multi-user tasks [12].

A natural application of future quantum networks is certainly the generalization of QKD to multiple users with multipartite QKD, also known as quantum conference key agreement (CKA) [19]. A CKA protocol is employed when a confidential message needs to be securely broadcast within a group of users. The users, upon performing a CKA protocol, share a common secret key—the conference key—with which they can encrypt and decrypt the secret message.

CKA plays an important role in this book. We introduce the reader to CKA by providing an intuitive explanation of its development from existing QKD protocols [20]. We generalize the security framework of QKD to include CKA and focus on a multipartite version of the popular BB84 protocol [21]. We also discuss the recent experimental realizations of CKA protocols, with particular emphasis on the implementation of the multipartite BB84 protocol [22].

Founded on the working principle of TF-QKD, we discuss a novel CKA protocol where multiple users distil a conference key through single-photon interference events [23]. Thanks to this feature, we show that the protocol significantly outperforms previous CKA schemes over long distances, as it employs a W-class state as its entanglement resource in place of the conventional GHZ state.

The information-theoretic security of QKD and CKA protocols holds as long as the assumptions made when proving their security are actually met by their experimental implementation. This requires the users to verify the trustworthiness of their quantum devices, which might be a quite daunting task.

A possible solution is provided by the device-independent (DI) paradigm. Indeed, the security of DI protocols, such as DIQKD and DI randomness generation (DIRG) protocols, holds independently of the actual functioning of the devices used to implement them [24, 25]. This remarkable fact relies on the observation of non-local correlations certified by a Bell inequality violation.

In this book we carefully review all the steps leading to the security of DI protocols starting from the observation of a Bell violation. In doing so, we revisit some well-established results in DIQKD [26] from the perspective of their recent generalization to multipartite DI protocols [27]. This allows us to introduce the latest developments in the security analyses of multiparty DI protocols and apply them to a specific tripartite DI scenario, as well as discuss the challenges in proving the security of DICKA protocols.

1.2 Book Structure

The contents of the book are organized as follows.

- In Chap. 2 we set the theoretical framework by introducing all the concepts of quantum information theory that are necessary for the understanding of the remainder of the book. We place particular emphasis on the various entropy definitions that capture different measures of information.
- We introduce quantum key distribution (QKD) in Chap. 3. After discussing the purely quantum features on which the security of QKD is based, we describe the paradigmatic BB84 protocol. We then consider a generic QKD protocol and prove its security under the most general circumstances. We conclude the Chapter by listing some important state-of-the-art QKD experiments.
- By generalizing the BB84 protocol to a multipartite scenario, in Chap. 4 we introduce multipartite QKD, also known as quantum conference key agreement (CKA).

We then describe the functioning of a generic CKA protocol and prove its security. Finally we describe the first experimental implementations of CKA protocols.

- In Chap. 5 we draw attention to the security threats posed by performing QKD with imperfect quantum devices and discuss the solutions proposed so far. Specifically, we present the decoy-state method to deal with sources emitting multiple photons. We also introduce the concept of measurement-device-independent QKD, whose security is independent of the trustworthiness of the measurement devices.

- The subject of Chap. 6 is the novel TF-QKD protocol, which applies the solutions to the security threats discussed in the previous Chapter. In this Chapter we also present recent fundamental bounds on the performance of any point-to-point QKD protocol. We introduce TF-QKD by describing its first version and the improved version that we investigate. We summarize the results of our investigation with the support of plots simulating the protocol's performance in realistic conditions. Insight is provided on the theoretical results that enable a practical performance assessment of TF-QKD. The last part of the Chapter is devoted to the discussion of the CKA protocol inspired by the founding idea of TF-QKD.

- We start Chap. 7 by proving Bell's theorem and introducing the concept of Bell inequality. We show that quantum correlations can violate Bell inequalities and clarify the relations between local, quantum, no-signaling and causal correlations. We then elucidate the link between the violation of a Bell inequality and the security of a device-independent (DI) QKD protocol. From there, we introduce the archetypal DIQKD protocol based on the violation of the Clauser-Horne-Shimony-Holt inequality, and prove its security. We then present recent theoretical results enabling similar security proofs for multipartite DI protocols. We conclude the Chapter by presenting a multipartite Bell inequality specifically designed to be employed in a DICKA protocol.

- Chapter 8 contains some concluding remarks and provides an outlook on future research directions related to the topics addressed in the book.

References

1. Brassard, G. (2005). Brief history of quantum cryptography: A personal perspective. *IEEE Information Theory Workshop on Theory and Practice in Information-Theoretic Security*, 19–23.
2. Bennett, C. H., Brassard, G., Breidbart, S., & Wiesner, S. (1983). Quantum cryptography, or unforgeable subway tokens. In Chaum, D., Rivest, R. L., & Sherman, A. T., (eds.) *Advances in Cryptology*, pp. 267–275. Boston, MA: Springer US.
3. Wiesner, S. (1983). Conjugate coding. *SIGACT News*, *15*(1), 78–88.
4. Bennett, C. H., & Brassard, G. (1984). Quantum cryptography: Public key distribution and coin tossing. In *Proceedings of IEEE International Conference on Computers, Systems and Signal Processing*, pp. 175 – 179.
5. Ekert, A. K. (1991). Quantum cryptography based on Bell's theorem. *Physical Review Letters*, *67*, 661–663.
6. Bruß, D. (1998). Optimal eavesdropping in quantum cryptography with six states. *Physical Review Letters*, *81*, 3018–3021.

7. Scarani, V., Bechmann-Pasquinucci, H., Cerf, N. J., Dušek, M., Lütkenhaus, N., & Peev, M. (2009). The security of practical quantum key distribution. *Reviews of Modern Physics, 81*, 1301–1350.

8. Diamanti, E., Lo, H.-K., Qi, B., & Yuan, Z. (2016). Practical challenges in quantum key distribution. *npj Quantum Information, 2*(1), 16025.

9. Commission, E. The quantum flagship. https://qt.eu.

10. Technology, I. Q. Quantum key distribution (qkd) markets: 2019-2028. https://www.insidequantumtechnology.com/product/quantum-key-distribution-qkd-markets-2019-2028

11. Kimble, H. J. (2008). The quantum internet. *Nature, 453*(7198), 1023–1030.

12. Wehner, S., Elkouss, D., & Hanson, R. (2018). Quantum internet: A vision for the road ahead. *Science, 362*(6412).

13. Broadbent, A., Fitzsimons, J., & Kashefi, E. (2009). Universal blind quantum computation. In *2009 50th Annual IEEE Symposium on Foundations of Computer Science*, pp. 517–526.

14. Fitzsimons, J. F. (2017). Private quantum computation: An introduction to blind quantum computing and related protocols. *npj Quantum Information, 3*(1), 23.

15. Lucamarini, M., Yuan, Z. L., Dynes, J. F., & Shields, A. J. (2018). Overcoming the rate-distance limit of quantum key distribution without quantum repeaters. *Nature, 557*(7705), 400–403.

16. Curty, M., Azuma, K., & Lo, H.-K. (2019). Simple security proof of twin-field type quantum key distribution protocol. *npj Quantum Information, 5*(1), 64.

17. Grasselli, F., & Curty, M. (2019). Practical decoy-state method for twin-field quantum key distribution. *New Journal of Physics, 21*(7), 073001.

18. Grasselli, F., Navarrete, Á., & Curty, M. (2019). Asymmetric twin-field quantum key distribution. *New Journal of Physics, 21*(11), 113032.

19. Murta, G., Grasselli, F., Kampermann, H., & Bruß, D. (2020). Quantum conference key agreement: A review. arXiv:quant-ph/2003.10186.

20. Epping, M., Kampermann, H., Macchiavello, C., & Bruß, D. (2017). Multi-partite entanglement can speed up quantum key distribution in networks. *New Journal of Physics, 19*(9), 093012.

21. Grasselli, F., Kampermann, H., & Bruß, D. (2018). Finite-key effects in multipartite quantum key distribution protocols. *New Journal of Physics, 20*(11), 113014.

22. Proietti, M., Ho, J., Grasselli, F., Barrow, P., Malik, M., & Fedrizzi, A. (2020). Experimental quantum conference key agreement. arXiv:quant-ph/2002.01491.

23. Grasselli, F., Kampermann, H., & Bruß, D. (2019). Conference key agreement with single-photon interference. *New Journal of Physics, 21*(12), 123002.

24. Yao, A., Mayers, D., & (1998). Quantum cryptography with imperfect apparatus. In *2013 IEEE 54th Annual Symposium on Foundations of Computer Science, p. 503, Los Alamitos, CA*. USA: IEEE Computer Society.

25. Acín, A., Gisin, N., & Masanes, L. (2006). From bell's theorem to secure quantum key distribution. *Physical Review Letters, 97*, 120405.

26. Pironio, S., Acín, A., Brunner, N., Gisin, N., Massar, S., & Scarani, V. (2009). Device-independent quantum key distribution secure against collective attacks. *New Journal of Physics, 11*(4), 045021.

27. Grasselli, F., Murta, G., Kampermann, H., & Bruß, D. (2020). Analytical entropic bounds for multiparty device-independent cryptography. arXiv:quant-ph/2004.14263.

Chapter 2
Elements of Quantum Information Theory

Fundamental measures of information arise as the answers to fundamental questions about the physical resources required to solve some information processing problem. Nielsen & Chuang

Abstract In this Chapter we review some fundamental concepts of quantum mechanics and linear algebra using the Dirac notation and the density operator formalism, including the notions of qubits, entanglement and general quantum operations (Sects. 2.1–2.5). We then introduce the entropies characterizing information-processing tasks which commonly occur in quantum cryptography, in particular the Shannon and von Neumann entropy (Sect. 2.6), the min- and max-entropy (Sect. 2.7) and their smooth versions (Sect. 2.8). We emphasize the operational meaning of the smooth min- and max-entropy for the cryptographic tasks of error correction and privacy amplification in Sects. 2.9 and 2.10, respectively. We additionally elaborate on the notion of distance between quantum states and its relation to their distinguishability in Sect. 2.11 in the Appendix of this Chapter. The content of this Chapter is mostly inspired by the following literature: [1–4].

2.1 Dirac Notation and Linear Algebra

The state of a quantum mechanical system, with d degrees of freedom, is represented by a normalized vector $|\psi\rangle$ in a d-dimensional Hilbert space $\mathcal{H}$ over the complex numbers $\mathbb{C}$, called the *state space* of the system. A *Hilbert space* is an inner product space, which is also complete with respect to the norm induced by the inner product if the space is infinite-dimensional.

The vector symbol $|\psi\rangle$ is called a *ket*. To every vector $|\psi\rangle$ in $\mathcal{H}$ corresponds a unique dual vector $\langle\psi|$ in the dual Hilbert space $\mathcal{H}^*$, i.e. the space of linear maps from $\mathcal{H}$ to $\mathbb{C}$. The symbol $\langle\phi|$ of a dual vector is called a *bra*. Note that the dual of a linear combination of vectors $\alpha|a\rangle + \beta|b\rangle$ is defined as $\alpha^*\langle a| + \beta^*\langle b|$, where α^* is the complex conjugate of $\alpha \in \mathbb{C}$.

The action of a linear map $\langle\phi| \in \mathcal{H}^*$ on a vector $|\psi\rangle \in \mathcal{H}$ is written as a "bra-ket": $|\psi\rangle \mapsto \langle\phi|\psi\rangle \in \mathbb{C}$ and defines the inner product of vectors $|\psi\rangle$ and $|\phi\rangle$ in $\mathcal{H}$. For the definition of dual vector, it follows that: $\langle\phi|\psi\rangle^* = \langle\psi|\phi\rangle$. Moreover, given a linear operator A, the quantity $\langle\phi|A|\psi\rangle \in \mathbb{C}$ can be interpreted as the result of the inner product between vectors $A|\psi\rangle$ and $|\phi\rangle$ in $\mathcal{H}$.

Two vectors are said to be *orthogonal* if their inner product is zero. The *norm* induced by the inner product is given by: $\||\psi\rangle\| = \sqrt{\langle\psi|\psi\rangle}$. A vector $|\psi\rangle$ is said to be *normalized*, or called a *unit vector*, if $\||\psi\rangle\| = 1$. An *orthonormal* set of vectors $\{|\psi_i\rangle\}$ is exclusively composed of normalized and mutually orthogonal vectors: $\langle\psi_i|\psi_j\rangle = \delta_{i,j}$, where $\delta_{i,j}$ is the Kronecker delta.

The Dirac notation provides a useful way to represent the action of linear operators on $\mathcal{H}$, through the *outer product*. The outer product of $|\psi\rangle \in \mathcal{H}$ and $\langle\phi| \in \mathcal{H}^*$ is represented by $|\psi\rangle\langle\phi|$ and acts as follows on $|\gamma\rangle \in \mathcal{H}$: $|\gamma\rangle \mapsto \langle\phi|\gamma\rangle|\psi\rangle$. The outer product of a vector $|\psi\rangle$ by itself defines a linear operator that projects a vector $|\phi\rangle \in \mathcal{H}$ in the one-dimensional subspace spanned by $|\psi\rangle$: $|\psi\rangle\langle\psi||\phi\rangle = \langle\psi|\phi\rangle|\psi\rangle$.

From this definition, it immediately follows that any orthonormal basis $\{|b_i\rangle\}_{i=1}^d$ of the d-dimensional Hilbert space $\mathcal{H}$ satisfies the *completeness relation*: $\sum_{i=1}^d |b_i\rangle\langle b_i| = \mathbb{1}$, where $\mathbb{1}$ is the identity operator. With the completeness relation, it is possible to represent the action of any linear operator A in the outer product notation:

$$A = \mathbb{1} A \mathbb{1} = \sum_{i,j=1}^d \langle b_i|A|b_j\rangle|b_i\rangle\langle b_j|, \qquad (2.1)$$

where the element $\langle b_i|A|b_j\rangle$ can be regarded as the matrix entry in the i-th row and j-th column of the *matrix representation* of A with respect to the basis $\{|b_i\rangle\}_{i=1}^d$.

Let A be a linear operator with domain $\mathcal{D}(A)$. We define the *adjoint* operator $A^\dagger$ of A as that operator with domain $\mathcal{D}(A^\dagger)$ such that:

$$(\langle\psi|A^\dagger|\phi\rangle)^* = \langle\phi|A|\psi\rangle \quad \forall|\psi\rangle \in \mathcal{D}(A), \ \forall|\phi\rangle \in \mathcal{D}(A^\dagger). \qquad (2.2)$$

This implies that the matrix representing $A^\dagger$ is obtained from that of A by applying transposition and complex conjugation. It also follows that $(|\psi\rangle\langle\phi|)^\dagger = |\phi\rangle\langle\psi|$ and that the dual vector of $A|\psi\rangle$ is $\langle\psi|A^\dagger$.

The evolution of a closed quantum system is determined by a *unitary* operator U, that is an operator for which $U^\dagger = U^{-1}$, where U^{-1} is the inverse of U. A unitary transformation also links any two bases $\{b_i\}_{i=1}^d$ and $\{b_i'\}_{i=1}^d$ in $\mathcal{H}$: $|b_i'\rangle = U|b_i\rangle$. Two bases are called *mutually unbiased* if $\langle b_i'|b_j\rangle = 1/d$ for every i and j.

The observable quantities in quantum mechanics are represented by *self-adjoint* operators. Let us take a moment to distinguish symmetric, self-adjoint and Hermitian operators.

A linear operator A with domain $\mathcal{D}(A)$ is called *symmetric* if

$$(\langle \psi | A | \phi \rangle)^* = \langle \phi | A | \psi \rangle \quad \forall | \psi \rangle, | \phi \rangle \in \mathcal{D}(A). \tag{2.3}$$

The above definition, combined with the definition of adjoint (2.2), implies that $\mathcal{D}(A) \subseteq \mathcal{D}(A^\dagger)$ and that A and $A^\dagger$ act in the same way on the vectors in $\mathcal{D}(A)$.

A linear operator A with domain $\mathcal{D}(A)$ is called *self-adjoint* if it is symmetric and $\mathcal{D}(A) = \mathcal{D}(A^\dagger)$. In other words, the operator A and its adjoint $A^\dagger$ coincide: $A = A^\dagger$.

Finally, a linear operator A is called *Hermitian* if it is self-adjoint ($A = A^\dagger$) and bounded. Note that a linear operator A is *bounded* if there exists a number $M \geq 0$ such that:

$$\| A | \psi \rangle \| \leq M \, \| | \psi \rangle \| \quad \forall | \psi \rangle \in \mathcal{H}. \tag{2.4}$$

A bounded operator is typically assumed to be defined on the whole Hilbert space $\mathcal{H}$ on which it acts. Therefore a linear operator is Hermitian if $A = A^\dagger$, $\mathcal{D}(A) = \mathcal{H}$ and (2.4) holds.

We remark that if $\mathcal{H}$ is finite-dimensional, then any linear operator A is bounded. This implies that for finite-dimensional Hilbert spaces the definitions of symmetric, self-adjoint and Hermitian operator are all equivalent.

In this book, unless otherwise specified, we always implicitly consider linear operators acting on finite-dimensional Hilbert spaces over $\mathbb{C}$.

An operator P is called a *projector* if $P^2 = P$. If P is also Hermitian, then it is called an *orthogonal* projector. Indeed, an orthogonal projector P and its complement, $\mathbb{1} - P$, project the same vector $|v\rangle$ onto orthogonal subspaces, as shown by the inner product of the two projected vectors:

$$\langle v | P^\dagger (\mathbb{1} - P) | v \rangle = \langle v | P (\mathbb{1} - P) | v \rangle = \langle v | (P - P^2) | v \rangle = 0. \tag{2.5}$$

An example of orthogonal projector is given by the following *rank-one* orthogonal projector $| \psi \rangle \langle \psi |$. Note that any orthogonal projector can be written as:

$$P = \sum_{i \in S} | b_i \rangle \langle b_i |, \quad S \subseteq \{1, \ldots, d\}, \tag{2.6}$$

i.e. as a sum of rank-one projectors on some elements of an orthonormal basis $\{| b_i \rangle\}_{i=1}^d \subset \mathcal{H}$.

Importantly, the eigenvalues a_i of an Hermitian operator $A = A^\dagger$ on the d-dimensional Hilbert space $\mathcal{H}$ are real and the eigenvectors form an orthonormal basis $\{|a_i\rangle\}_{i=1}^d$, called the *eigenbasis* of A. Then, the operator A can be written in its *spectral decomposition* as follows[1]:

$$A = \sum_{i=1}^d a_i |a_i\rangle\langle a_i|, \tag{2.7}$$

where $|a_i\rangle\langle a_i|$ are rank-one projectors on the corresponding eigenvectors $|a_i\rangle$.

We define the *trace* of an operator A on $\mathcal{H}$ as follows:

$$\mathrm{Tr}[A] = \sum_{i=1}^d \langle b_i | A | b_i \rangle, \tag{2.8}$$

where $\{|b_i\rangle\}_{i=1}^d$ is any orthonormal basis for $\mathcal{H}$. We remark that the trace definition is independent of the chosen basis thanks to the cyclic property of the trace $\mathrm{Tr}[AB] = \mathrm{Tr}[BA]$ and to the fact that a change of orthonormal basis is represented by a unitary operator.

At last, an operator A is said to be *positive* $(A \geq 0)$ if its expectation value on any vector is non-negative: $\langle v | A | v \rangle \geq 0$ for all $|v\rangle \in \mathcal{H}$. Notably, an operator A is positive if and only if it is Hermitian $(A = A^\dagger)$ with non-negative eigenvalues.

2.2 Density Operator Formalism

We have so far identified the state of a quantum system by its wave function $|\psi\rangle$, implicitly assuming that it can be completely determined. However, from a practical point of view, this is not always feasible. Consider, for instance, an electron-target scattering experiment where the electron beam is prepared without the use of polarizers. The electron spin will probably be oriented in a random direction for each electron of the beam. Thus, the spin of the beam cannot be described by a *pure state* of the form:

$$|\psi\rangle = \alpha |\uparrow\rangle_z + \beta |\downarrow\rangle_z, \tag{2.9}$$

since the latter describes a spin oriented in a specific direction, fixed by the polar angles $\theta = 2\arccos|\alpha|$ and $\varphi = \arg\beta - \arg\alpha$. Rather, the state of the beam spin is described by an ensemble of spins oriented in all directions, weighted by their probability of occurrence: a *mixed state*.

In cases like this, where the lack of information on an ensemble of single states prevents us from describing them one by one completely, we can still study such a collection of states statistically by means of the *density operator*, introduced by von Neumann in 1927.

---

[1]Note that the spectral theorem also holds for self-adjoint operators, but does not hold for merely symmetric operators.

Definition 2.1 (*Density Operator*) Consider a quantum system that is found in one of the pure states $\{|\psi_i\rangle\}$ with probabilities $p_i < 1$, where $\sum_i p_i = 1$. Then, the state of the system is called a mixed state and is described by the density operator

$$\rho = \sum_i p_i |\psi_i\rangle\langle\psi_i|. \tag{2.10}$$

The matrix representation of ρ is called density matrix, which is often used to indicate the operator itself.

If a quantum system is in a pure state $|\psi\rangle$ with certainty, then its density operator reads: $\rho = |\psi\rangle\langle\psi|$. A criterion to determine whether a state ρ is pure or mixed is given by the computation of its *purity*: $\mathrm{Tr}[\rho^2]$. A state is pure if $\mathrm{Tr}[\rho^2] = 1$, while it's mixed if $\mathrm{Tr}[\rho^2] < 1$.

Density operators offer an alternative formulation of quantum mechanics, which is particularly useful in quantum information. Here we provide an intrinsic characterization of density operators, which allows us to abandon their interpretation in terms of an ensemble of pure states.

Theorem 2.1 (Characterization of density operators) *An operator ρ is the density operator of a mixed state $\rho = \sum_i p_i |\psi_i\rangle\langle\psi_i|$ if and only if it is normalized (*$\mathrm{Tr}[\rho] = 1$*) and positive.*

The proof of this Theorem can be found in [2].

We can now reformulate the postulates of quantum mechanics in the density operator picture.

Postulate 2.1 *The state of a quantum system is completely determined by a normalized positive operator, denoted density operator, acting on a Hilbert space $\mathcal{H}$ named the state space of the system.*

Postulate 2.2 *The evolution of a closed quantum system is determined by a unitary transformation U. Specifically, the evolved state of the system ρ' is obtained from the initial state ρ as follows:*

$$\rho' = U\rho U^\dagger. \tag{2.11}$$

Postulate 2.3 *The measurement of a quantum system is defined by a collection of measurement operators $\{M_m\}$ acting on $\mathcal{H}$ and satisfying the completeness relation: $\sum_m M_m^\dagger M_m = \mathbb{1}$. If ρ is the state of the system prior to measurement, the probability of observing the measurement outcome m is given by:*

$$\Pr(m) = \mathrm{Tr}[M_m^\dagger M_m \rho] \tag{2.12}$$

and the state of the system after the measurement reads

$$\rho_m = \frac{M_m \rho M_m^\dagger}{\mathrm{Tr}[M_m^\dagger M_m \rho]}. \tag{2.13}$$

Postulate 2.4 *The state space $\mathcal{H}$ of a composite quantum system, with subsystems numbered from 1 to N, is given by the tensor product of the state spaces $\mathcal{H}_i$ composing the system: $\mathcal{H} = \mathcal{H}_1 \otimes \mathcal{H}_2 \otimes \cdots \otimes \mathcal{H}_N$.*

Thanks to Postulates 2.1 and 2.4, we can describe the state of a composite system commonly encountered in quantum information. Consider a quantum system Q whose state depends on the value x of a classical random variable X, with probability distribution $\Pr(x)$. For an observer who ignores the value of X, the global state of the quantum system and of the classical variable is given by:

$$\rho_{XQ} = \sum_x \Pr(x) |x\rangle\langle x|_X \otimes \rho_Q^x, \qquad (2.14)$$

where the random variable X is represented by orthogonal pure states $|x\rangle$, since its classical outcomes can be perfectly distinguished. The quantum system is instead found in one of the *conditional states* ρ_Q^x. Moreover, we say that ρ_{XQ} is *classical on X* or is a *classical-quantum* (c.q) state if it can be written in the form (2.14).

2.2.1 POVMs and Projective Measurements

Postulate 2.3 provides the most general description of a quantum measurement. There are two special cases of quantum measurements which are of particular interest in quantum information. The first one is the positive operator-valued measure (POVM), which simplifies the formalism when only the measurement statistics matters.

Definition 2.2 *(POVM)* A POVM is defined by a set of positive operators $\{E_m\}$, the POVM elements, acting on the state space, such that $\sum_m E_m = \mathbb{1}$. Then the probability of obtaining outcome m when measuring the system in state ρ is given by:

$$\Pr(m) = \mathrm{Tr}[E_m \rho]. \qquad (2.15)$$

One can readily see that POVMs are a special case of Postulate 2.3, when the measurement operators are given by $M_m = \sqrt{E_m}$, which implies $M_m^\dagger M_m = E_m$.

The only case in which the measurement operators and the POVM elements coincide is for *projective measurements*, i.e. when they are orthogonal projectors: $E_m = M_m = P_m$. One can verify that a set of orthogonal projectors $\{P_m\}$, satisfying the completeness relation: $\sum_m P_m = \mathbb{1}$, is composed of mutually orthogonal projectors: $P_m P_n = \delta_{m,n} P_n$ [5]. From the measurement outcomes and the projectors it is possible to define an Hermitian operator M, called *observable*, through its spectral decomposition:

$$M = \sum_m m P_m. \qquad (2.16)$$

Often, a projective measurement is equivalently defined as the measurement of an observable M, meaning that the projectors are those appearing in its spectral decomposition (2.16). If the projectors are all rank-one $P_m = |m\rangle\langle m|$, the measurement is called a *von Neumann measurement*.

Identifying a projective measurement with the observable M is useful when, for instance, one wants to compute the average outcome, since it can be directly written in terms of the observable M:

$$\langle M \rangle := \sum_m m \Pr(m) = \sum_m m \, \text{Tr}[P_m \rho] = \text{Tr}[M\rho]. \tag{2.17}$$

Finally we remark that, although projective measurements are particular cases of POVMs, the statistics of any POVM on a d-dimensional Hilbert space can be reproduced by combining a projective measurement on a Hilbert space of dimension $d' \geq d$ with a unitary operation. This result is known as the Naimark theorem [6, 7].

2.3 Qubits and Pauli Operators

In many quantum information applications, the fundamental quantum system is a two-level system called quantum bit or *qubit*. Physical realizations of qubits are, for example: a photon that can be found in one of two distinct paths, two orthogonal polarizations of a photon, the spin state of spin-$\frac{1}{2}$ particles, or the two lowest energy levels of an electron orbiting a nucleus.

The state space of a qubit is a two-dimensional Hilbert space, $\mathcal{H}_2$. The commonly used basis for $\mathcal{H}_2$ is the *computational basis* $\{|0\rangle, |1\rangle\}$. Thus, any pure qubit state is represented by a superposition of the form: $|\psi\rangle = \alpha|0\rangle + \beta|1\rangle$, where $|\alpha|^2 + |\beta|^2 = 1$.

Conversely, any mixed qubit state is represented by a density operator ρ acting on $\mathcal{H}_2$ and can be expressed as a combination of the identity operator $\mathbb{1}$ and the Pauli operators σ_x, σ_y and σ_z [8]:

$$\rho = \frac{\mathbb{1} + \mathbf{r} \cdot \boldsymbol{\sigma}}{2}, \quad \mathbf{r} \in \mathbb{R}^3 : \|r\| \leq 1, \tag{2.18}$$

where $\boldsymbol{\sigma} = (\sigma_x, \sigma_y, \sigma_z)^T$. The matrix representation of the Pauli operators with respect to the computational basis reads:

$$\sigma_x = \begin{bmatrix} 0 & 1 \\ 1 & 0 \end{bmatrix} ; \quad \sigma_y = \begin{bmatrix} 0 & -i \\ i & 0 \end{bmatrix} ; \quad \sigma_z = \begin{bmatrix} 1 & 0 \\ 0 & -1 \end{bmatrix} \tag{2.19}$$

Note that, depending on the context, we also indicate the Pauli operators as $\sigma_x = \sigma_1 = X$, $\sigma_y = \sigma_2 = Y$ and $\sigma_z = \sigma_3 = Z$. The Pauli operators are Hermitian with eigenvalues ± 1, traceless, and satisfy the following relation:

$$\sigma_i \sigma_j = \delta_{i,j} \mathbb{1} + \sum_{k=1}^{3} \varepsilon_{ijk} \sigma_k, \qquad (2.20)$$

where ε_{ijk} is the Levi-Civita symbol, which is equal to $+1$ (or -1) if the triple (i, j, k) is a cyclic (or anti-cyclic) permutation of $(1, 2, 3)$, and zero if any two indices are repeated.

The representation (2.18) of a qubit state, together with (2.20), is particularly useful in many computations. For instance, the purity of a qubit state can be readily computed as: $\mathrm{Tr}[\rho^2] = (1 + \|r\|^2)/2$. Thus, the norm of the vector $\mathbf{r}$ indicates whether the state is pure ($\|r\| = 1$) or mixed ($\|r\| < 1$). When $\|r\| = 0$, the state is said to be *maximally mixed*.

Moreover, the vector $\mathbf{r}$ individuates a point inside a unit sphere, called the *Bloch sphere*, where the three Cartesian coordinates are associated with the eigenstates of the Pauli operators. Often, in the quantum information jargon one can measure "in the z direction of the Bloch sphere", meaning that one is performing a projective measurement in the eigenbasis of σ_z, which is conventionally associated with the computational basis.

Finally, the Pauli operators have a prominent role in quantum error correction, since they represent all the possible errors that can occur when processing a qubit. In particular, σ_x produces bit flips, σ_z yields phase flips and σ_y both phase and bit flips:

$$\sigma_x |a\rangle = |\bar{a}\rangle \qquad (2.21)$$
$$\sigma_z |a\rangle = (-1)^a |a\rangle \qquad (2.22)$$
$$\sigma_y |a\rangle = \mathrm{i}(-1)^a |\bar{a}\rangle, \quad a = 0, 1 \qquad (2.23)$$

where $\bar{a} = 1 - a$.

2.4 Composite Systems and Entanglement

Postulate 2.4 allows us to introduce one of the most astonishing features of quantum mechanics, entanglement, which plays a crucial role in quantum information protocols.

Suppose that two parties, Alice and Bob, locally prepare their own quantum system in the pure states $|\psi_A\rangle$ and $|\psi_B\rangle$, respectively. Then, by Postulate 2.4, the state of the composite quantum system is pure: $|\Psi\rangle = |\psi_A\rangle \otimes |\psi_B\rangle$ and is called a *product state*. Note that a compact notation for $|\psi_A\rangle \otimes |\psi_B\rangle$ is $|\psi_A, \psi_B\rangle$ or $|\psi_A \psi_B\rangle$. If the pure state of a composite system is not a product state, then it is called *entangled*.

More generally, Alice and Bob could agree on locally preparing the states $|\psi_A^i\rangle$ and $|\psi_B^i\rangle$ according to the value of a shared random variable with distribution $\{p_i\}$. This task only requires local operations and classical communication (LOCC). In this case, the state of the composite system is described by the mixed state:

$$\rho_{AB} = \sum_i p_i \, |\psi_A^i\rangle\langle\psi_A^i| \otimes |\psi_B^i\rangle\langle\psi_B^i|$$

$$= \sum_i p_i \, |\psi_A^i, \psi_B^i\rangle\langle\psi_A^i, \psi_B^i| \tag{2.24}$$

and is called a *separable state*. Note that product states are a particular case of separable states and that the state (2.24) is the most general state obtained by LOCC. Indeed, even if Alice and Bob prepare their systems in mixed states ρ_A^i and ρ_B^i, the state of the composite system, $\sum_i p_i \, \rho_{A_i} \otimes \rho_{B_i}$, can always be reduced to one of the form (2.24).

However, not every composite quantum system is prepared with LOCC, hence its state may not be a separable state (2.24) [9]. In that case, we say that the state of the composite system is entangled.

Definition 2.3 (*Separability, Entanglement*) A quantum state ρ_{AB} on $\mathcal{H}_A \otimes \mathcal{H}_B$ is called separable if there exists a convex combination of pure product states $|\psi_A^i\rangle \otimes |\psi_B^i\rangle$, with $|\psi_A^i\rangle \in \mathcal{H}_A$ and $|\psi_B^i\rangle \in \mathcal{H}_B$, such that:

$$\rho_{AB} = \sum_i p_i \, |\psi_A^i, \psi_B^i\rangle\langle\psi_A^i, \psi_B^i|. \tag{2.25}$$

Otherwise, ρ_{AB} is called entangled.

Classifying whether a state is entangled or not is challenging. Consider for example the following pure states, known as *Bell states*:

$$|\Phi^{\pm}\rangle = \frac{|00\rangle \pm |11\rangle}{\sqrt{2}}, \tag{2.26}$$

whose density operators are clearly entangled according to Definition 2.3:

$$|\Phi^{\pm}\rangle\langle\Phi^{\pm}| = \frac{1}{2}\left[|00\rangle\langle00| \pm |00\rangle\langle11| \pm |11\rangle\langle00| + |11\rangle\langle11|\right]. \tag{2.27}$$

Surprisingly, their convex combination is not entangled:

$$\frac{1}{2}\left[|\Phi^{+}\rangle\langle\Phi^{+}| + |\Phi^{-}\rangle\langle\Phi^{-}|\right] = \frac{1}{2}\left[|00\rangle\langle00| + |11\rangle\langle11|\right]. \tag{2.28}$$

The definition of entanglement can be extended to a multipartite scenario. An N-partite pure state is called fully separable if one can assign a single state vector to the subsystem of each party, i.e. if it is a product state: $|\psi_A^1\rangle \otimes \cdots \otimes |\psi_A^N\rangle$. Otherwise the state is entangled. The more general definition valid for mixed states is the following.

Definition 2.4 (*Multipartite Entanglement*, [10]) Consider a set of N parties labelled by the indices $\mathcal{I} = \{1, 2, \ldots, N\}$ and a partition $\{\mathcal{I}_j\}_{j=1}^k$ of $\mathcal{I}$, where $\mathcal{I}_j$ are disjoint subsets of $\mathcal{I}$ such that $\cup_j \mathcal{I}_j = \mathcal{I}$. Then a state ρ is k-separable with respect to the partition $\{\mathcal{I}_j\}_{j=1}^k$ if it can be written as a convex combination of product states on the k partitions:

$$\rho = \sum_i p_i \, \rho_{\mathcal{I}_1}^i \otimes \cdots \otimes \rho_{\mathcal{I}_k}^i. \tag{2.29}$$

If every $\mathcal{I}_j$ comprises only one index, then the state (2.29) is called fully separable. If a state is not fully separable, then it is entangled. If the partition is only composed of two subsets $\mathcal{I}_1$ and $\mathcal{I}_2$, the state (2.29) is called biseparable. If a state cannot be expressed as a convex combination of biseparable states, then it is called genuine multipartite entangled (GME).

An example of a pure GME state which plays a major role in multipartite quantum cryptographic protocols is the Greenberger–Horne–Zeilinger (GHZ) state [11]:

$$|\text{GHZ}_N\rangle = \frac{1}{\sqrt{2}} \left[|0\rangle^{\otimes N} + |1\rangle^{\otimes N} \right]. \tag{2.30}$$

A remarkable application of the density operator formalism is the ability to describe subsystems of composite systems through the *reduced density operator*. This is particularly useful when the global state is entangled and the states of its subsystems are not immediately intelligible.

Definition 2.5 (*Reduced density operator*) Let ρ_{AB} be the state of a bipartite quantum system. Then the reduced density operator representing the state on subsystem A is given by:

$$\rho_A = \text{Tr}_B[\rho_{AB}], \tag{2.31}$$

where Tr_B is the partial trace on subsystem B.

The partial trace is defined as the regular trace (2.8) but only acts on the subsystems indicated in the subscript. For instance, given two orthonormal bases $\{|a_i\rangle\}_i$ and $\{|b_j\rangle\}_j$ for the two subsystems A and B, we can express ρ_{AB} in the outer product notation and compute the action of the partial trace as follows:

$$\begin{aligned} \text{Tr}_B[\rho_{AB}] &= \sum_{i,j,k,l} r_{(i,j),(k,l)} \, \text{Tr}_B[|a_i\rangle\langle a_k| \otimes |b_j\rangle\langle b_l|] \\ &= \sum_{i,j,k,l} r_{(i,j),(k,l)} \langle b_l|b_j\rangle |a_i\rangle\langle a_k|, \end{aligned} \tag{2.32}$$

where $r_{(i,j),(k,l)} = \langle a_i, b_j|\rho_{AB}|a_k, b_l\rangle$ are the matrix elements of ρ_{AB}.

Definition 2.5 is justified by the fact that the reduced density operator ρ_A provides the correct measurement statistics for measurements made on subsystem A.

2.4.1 The Schmidt Decomposition and Purifications

In Sect. 2.4 we observed that entanglement detection is not an easy task and that a pure state is entangled if it cannot be expressed as a product state. A useful tool to detect entanglement in bipartite pure states is given by the *Schmidt decomposition*.

Theorem 2.2 (Schmidt decomposition) *Let* $|\psi_{AB}\rangle \in \mathcal{H}_A \otimes \mathcal{H}_B$ *be the pure state of a bipartite system. Then there exists an orthonormal basis* $\{|\alpha_i\rangle\}_{i=1}^{d_A}$ *of* $\mathcal{H}_A$ *and an orthonormal basis* $\{|\beta_j\rangle\}_{j=1}^{d_B}$ *of* $\mathcal{H}_B$ *such that:*

$$|\psi_{AB}\rangle = \sum_{k=1}^{R} \lambda_k |\alpha_k, \beta_k\rangle, \tag{2.33}$$

where λ_k *are positive real coefficients called Schmidt coefficients and* $R \leq \min(d_A, d_B)$ *is the Schmidt rank.*

Proof Let $\{|a_i\rangle\}_{i=1}^{d_A}$ and $\{|b_j\rangle\}_{j=1}^{d_B}$ two orthonormal bases of $\mathcal{H}_A$ and $\mathcal{H}_B$, respectively. Then the state $|\psi_{AB}\rangle$ can be expressed as:

$$|\psi_{AB}\rangle = \sum_{i,j=1}^{d_A, d_B} c_{ij} |a_i, b_j\rangle, \tag{2.34}$$

for some complex coefficients c_{ij} which define the complex matrix $C \in \mathbb{C}^{d_A \times d_B}$. From the singular value decomposition of C we obtain: $C = UDV$, where $U \in \mathbb{C}^{d_A \times d_A}$ and $V \in \mathbb{C}^{d_B \times d_B}$ are unitary matrices and $D \in \mathbb{R}^{d_A \times d_B}$ is a rectangular diagonal matrix of non-negative numbers, the singular values of C. By substituting the expression for C in (2.34) we get:

$$|\psi_{AB}\rangle = \sum_{k=1}^{\min(d_A, d_B)} \sum_{i,j=1}^{d_A, d_B} u_{ik} d_{kk} v_{kj} |a_i, b_j\rangle, \tag{2.35}$$

where u_{ik}, d_{kk} and v_{kj} are the matrix elements of U, D and V, respectively. We now define new basis elements $|\alpha_k\rangle = \sum_{i=1}^{d_A} u_{ik} |a_i\rangle$ and $|\beta_k\rangle = \sum_{j=1}^{d_B} v_{kj} |b_j\rangle$. The newly defined bases $\{\alpha_k\}_{k=1}^{d_A}$ and $\{\beta_k\}_{k=1}^{d_B}$ are orthonormal since the starting ones were so. By substituting the bases in (2.35) and by discarding the terms in the sum over k where $d_{kk} = 0$, we obtain the claim in (2.33). $\square$

Note that the Schmidt coefficients are given by the non-zero singular values of C, which can be computed as the square roots of the non-zero eigenvalues of $CC^\dagger$.

The Schmidt decomposition allows us to immediately compute the reduced state of the two subsystems according to Definition 2.5:

$$\rho_A = \sum_{k=1}^{R} \lambda_k^2 |\alpha_k\rangle\langle\alpha_k| \quad ; \quad \rho_B = \sum_{k=1}^{R} \lambda_k^2 |\beta_k\rangle\langle\beta_k| \tag{2.36}$$

We observe that ρ_A and ρ_B are written in their spectral decomposition and have the same eigenvalues! Since many properties in quantum information are determined by the eigenvalues of a state (e.g., the von Neumann entropy, see Sect. 2.6), they will be the same for the two subsystems of a composite quantum system in a pure state.

Moreover, if the Schmidt rank is $R = 1$, the global state is separable and the two subsystems are pure. Otherwise, for $R > 1$ the global state is entangled and the two subsystems are mixed. In this case, we can completely determine the (pure) state of the combined system (2.33), but we lack information when we focus on its single constituents (2.36)—the reduced states are mixed. This bizarre fact is one of the hallmarks of entanglement.

The missing information on system A is represented by its classical randomness (2.36), which is correlated to system B as visualized in (2.33). Only a global description of systems A and B, provided by the pure state (2.33), presents no classical randomness and hence cannot be correlated with any other system. Therefore, everything that might possibly be correlated with system A is contained in system B.

This fact is widely used in quantum cryptography. Here, a group of honest parties holds a quantum system A. One then assumes the worst-case scenario where the eavesdropper, Eve, holds the quantum system E that contains all the possible correlations with A, i.e. the composite system AE is in a pure state. We say that Eve holds the *purifying system*, which can be identified as follows.

Proposition 2.1 (Purification) *Let ρ_A on $\mathcal{H}_A$ be the state of a quantum system A. Then there exists an auxiliary system E with state space $\mathcal{H}_E$ and a pure state $|\psi_{AE}\rangle \in \mathcal{H}_A \otimes \mathcal{H}_E$, called a purification of ρ_A, such that:*

$$\mathrm{Tr}_E[|\psi_{AE}\rangle\langle\psi_{AE}|] = \rho_A. \tag{2.37}$$

Proof Consider the spectral decomposition of ρ_A: $\sum_{i=1}^{d} \lambda_i |\lambda_i\rangle\langle\lambda_i|$ and a Hilbert space $\mathcal{H}_E$ of the same dimension d of $\mathcal{H}_A$, with orthonormal basis $\{|e_i\rangle\}$. The purification of ρ_A is given by:

$$|\psi_{AE}\rangle = \sum_{i=1}^{d} \sqrt{\lambda_i} |\lambda_i\rangle \otimes |e_i\rangle. \tag{2.38}$$

Indeed, we have that:

$$\mathrm{Tr}_E[|\psi_{AE}\rangle\langle\psi_{AE}|] = \sum_{i,j=1}^{d} \sqrt{\lambda_i \lambda_j} |\lambda_i\rangle\langle\lambda_j| \, \mathrm{Tr}[|e_i\rangle\langle e_j|]$$

$$= \sum_{i=1}^{d} \lambda_i |\lambda_i\rangle\langle\lambda_i| = \rho_A, \qquad (2.39)$$

as claimed.   □

Note that all purifications $|\psi_{AE}\rangle$ of ρ_A are related by unitaries on E.

2.5 Quantum Operations

A quantum operation $\mathcal{E}$, also called quantum channel, provides the most general description of a physical process acting on a system in state ρ. The final state of the system, after the process occurs, is given by $\mathcal{E}(\rho)$ up to some normalization factor. Both the unitary evolution of a closed system (Postulate 2.2) and quantum measurements (Postulate 2.3) are examples of quantum operations.

Quantum operations are defined by the following three axiomatic properties, based on physical grounds.

Axiom 2.1 *The probability that the process represented by $\mathcal{E}$ occurs is given by* $\mathrm{Tr}[\mathcal{E}(\rho)] \in [0, 1]$*, when ρ is the initial state.*

Axiom 2.2 *The map $\mathcal{E}$ is convex-linear on the set of density operators, i.e.*

$$\mathcal{E}\left(\sum_i p_i \rho_i\right) = \sum_i p_i \mathcal{E}(\rho_i).$$

Axiom 2.3 *The map $\mathcal{E}$ is completely positive (CP). That is, $\mathcal{E}(\rho)$ is a positive operator for every input state ρ. Additionally, for every composite state ρ_{AB} on $\mathcal{H}_A \otimes \mathcal{H}_B$, the operator $(\mathbb{1}_A \otimes \mathcal{E})(\rho_{AB})$ is positive on $\mathcal{H}_A \otimes \mathcal{H}_B$.*

The axioms are chosen such that quantum operations map density operators to density operators. Axiom 2.1 includes quantum measurements (where each outcome occurs with a certain probability) as a possible quantum operation. The normalized state after the process in this case reads $\mathcal{E}(\rho)/\mathrm{Tr}[\mathcal{E}(\rho)]$. The second axiom states a desirable property, namely that if a system is in one of the states $\{\rho_i\}$ with distribution $\{p_i\}$, after applying $\mathcal{E}$ it will be in one of the states $\{\mathcal{E}(\rho_i)\}$ with the same probability distribution. Finally, Axiom 2.3 ensures that the output of a quantum operation is still a density operator, even when it acts on a subsystem of a composite system. Note that this requirement is non-trivial, as there are maps which are positive but not CP.

Kraus' theorem is a beautiful result which characterizes quantum operations with an elegant notation.

Theorem 2.3 (Kraus). *A map $\mathcal{E}$ is a quantum operation satisfying Axioms 2.1, 2.2 and 2.3 if and only if it can be represented by a Kraus decomposition:*

$$\mathcal{E}(\rho) = \sum_i K_i \rho K_i^\dagger, \tag{2.40}$$

for some set of Kraus operators $\{K_i\}$ such that $\sum_i K_i^\dagger K_i \leq \mathbb{1}$. Moreover, being d the dimension of the Hilbert space of the system on which $\mathcal{E}$ acts, the number of Kraus operators is not larger than d^2.

The proof of theorem 2.3 can be found in [2, 7]. We point out that the Kraus decomposition of a quantum operation is not unique, but all the possible decompositions are linked by unitary transformations.

Unitaries and measurements are two particular cases of quantum operations with one Kraus operator each, given by: $K = U$ and $K = M_m$, respectively. However, while unitaries are trace-preserving operations, $\mathrm{Tr}[U\rho U^\dagger] = \mathrm{Tr}[\rho U^\dagger U] = 1$, quantum measurements in general are not.

An equivalent description of quantum operations interprets them as the result of the interaction between the system of interest (S) and an environment (E). Conversely, in absence of interactions with an environment, the system would evolve according to a unitary transformation (Postulate 2.2). Suppose that the system and the environment are initially in a product state, where the environment is described by a pure state $|e_0\rangle$ and the system's state is ρ. Note that assuming an initial pure state for the environment is not restrictive as we did not fix its dimension, thus we could always take its purification. The composite system ($S + E$) is closed and evolves according to a unitary U. Then, the final state of system S reads:

$$\mathcal{E}(\rho) = \mathrm{Tr}_E\left[U(\rho \otimes |e_0\rangle\langle e_0|)U^\dagger\right] = \sum_i \langle e_i|U(\rho \otimes |e_0\rangle\langle e_0|)U^\dagger|e_i\rangle, \tag{2.41}$$

for some orthonormal basis $\{|e_i\rangle\}$ of the environment. By comparing (2.41) with (2.40), we deduce an explicit expression for the Kraus operators:

$$K_i = \langle e_i|U|e_0\rangle. \tag{2.42}$$

Since the operators (2.42) satisfy the completeness relation $\sum_i K_i^\dagger K_i = \mathbb{1}$, they describe trace-preserving quantum operations. Instead, non-trace-preserving quantum operations can be viewed as just described with an additional projective measurement on the combined system, following the unitary U.

2.5.1 Depolarizing Channel

An important example of quantum operation is the *depolarizing channel*:

$$\mathcal{E}(\rho) = (1 - p)\rho + \frac{p}{3} \sum_{i=1}^{3} \sigma_i \rho \sigma_i^\dagger, \tag{2.43}$$

with Kraus operators $K_0 = \sqrt{1 - p}\,\mathbb{1}$ and $K_i = \sqrt{p/3}\,\sigma_i$. Recall from Sect. 2.3 that every qubit error can be reproduced by applying a Pauli operator σ_i. Thus, the resulting state in (2.43) is unchanged with probability $1 - p$ or is affected by one of the qubit errors with probability $p/3$ each. By applying the depolarizing channel on the qubits used in a quantum information protocol, one can test the protocol's robustness against noise.

Note that, under the substitution $p = 3q/4$, the map in (2.43) can be recast as:

$$\mathcal{E}(\rho) = (1 - q)\rho + q\frac{\mathbb{1}}{2}, \tag{2.44}$$

i.e. it depolarizes a qubit with probability q by replacing it with the completely mixed state $\mathbb{1}/2$.

2.6 Shannon and von Neumann Entropy

The uncertainty that an observer has about a physical system, i.e. the amount of randomness characterizing the system from her perspective, is quantified by a certain entropy measure.

Definition 2.6 (*Shannon entropy*) Let X be a random variable whose outcomes follow the probability distribution $\{p_x\}$. The Shannon entropy of X (or of the distribution $\{p_x\}$) is given by:

$$H(X) = H(\{p_x\}) = -\sum_x p_x \log p_x. \tag{2.45}$$

> In this book the logarithm symbol "log" is always intended in base 2 and by convention it holds: $0 \log 0 = 0$.

The Shannon entropy quantifies the uncertainty about X before we learn its value. Equivalently, $H(X)$ are the bits of information gained after reading the outcome of X. If X has only two possible outcomes, the Shannon entropy is often called *binary entropy* and reads:

$$h(p) := -p \log p - (1 - p) \log(1 - p) \tag{2.46}$$

Given two random variables X and Y jointly distributed according to $\{p(x, y)\}$, the joint Shannon entropy reads:

$$H(XY) = -p(x, y) \sum_{x,y} p(x, y), \tag{2.47}$$

while the *conditional entropy* is defined as:

$$H(X|Y) = H(XY) - H(Y). \tag{2.48}$$

The conditional entropy of X given Y quantifies how uncertain we are about X, given that we learned the value of Y. Finally, the *mutual information* $H(X : Y)$ measures the amount of information we gain on X by observing the value of Y. This is given by the total amount of information of X, $H(X)$, minus the uncertainty that we still have on X after learning Y, i.e. $H(X|Y)$. Thus we have:

$$H(X : Y) = H(X) - H(X|Y) = H(X) + H(Y) - H(XY). \tag{2.49}$$

Of the many properties satisfied by the above-defined entropies, we highlight in particular that: $H(X|Y) = H(XY) - H(Y) \geq 0$. We could intuitively expect this, since the uncertainty on both random variables X and Y must be greater than the uncertainty on Y. Conversely, this does not hold in general for quantum states, whose uncertainty is quantified by the *von Neumann entropy*.

Definition 2.7 (*von Neumann entropy*) The von Neumann entropy of a quantum state ρ, with eigenvalues $\{\lambda_i\}$, is defined as:

$$H(\rho) = -\operatorname{Tr}[\rho \log \rho] = -\sum_i \lambda_i \log \lambda_i. \tag{2.50}$$

Often, the von Neumann entropy of a system A in state ρ is indicated as: $H(A)_\rho$.

One can interpret the von Neumann entropy of ρ as the Shannon entropy of the probability distribution defined by its eigenvalues, hence we use the same symbol. For this analogy, the previous definitions of joint entropy (2.47), conditional entropy (2.48) and mutual information (2.49) can be extended to the von Neumann entropy.

We observe that $0 \leq H(\rho) \leq \log d$ for every state ρ on a d-dimensional Hilbert space. Moreover $H(\rho) = 0$ if ρ is pure and $H(\rho) = \log d$ if the state is maximally mixed.

Suppose that the state ρ_{AB} of a composite system is given by the pure entangled state in (2.26), already written in its Schmidt decomposition. Then, the von Neumann entropy of the composite system is: $H(AB)_\rho = 0$. From the Schmidt decomposition (c.f. Sect. 2.4) we learned that the eigenvalues of ρ_A and ρ_B are equal and given by the squares of the Schmidt coefficients. This leads to: $H(A)_\rho = H(B)_\rho = 1$.

Clearly, in the quantum case the entropy of a subsystem can be larger than the entropy of the composite system and the conditional entropy becomes negative: $H(A|B) = H(AB) - H(B) = -1 < 0$.

Other important properties of the von Neumann entropy are the *strong subadditivity*:

$$H(A|BC) \leq H(A|B), \tag{2.51}$$

and the conditional entropy of a c.q. state. Let $\rho_{AB} = \sum_a \Pr(a) |a\rangle\langle a| \otimes \rho_B^a$ be a c.q. state, where the state on B depends on the value a. Then the entropy of B conditioned on A can be expressed as:

$$H(B|A)_\rho = \sum_a \Pr(a) H(\rho_B^a). \tag{2.52}$$

2.6.1 Operational Meaning

We conclude this Section by briefly providing an operational meaning of the Shannon and von Neumann entropies, which helps us motivate the introduction of smooth entropies.

Consider a source emitting a sequence of random symbols represented by random variables $X_1, X_2,...,X_n$, each of them distributed according to P_X and independent from each other. They are said to be *independent and identically distributed* (i.i.d.) random variables. The goal is to store the data by encoding it in a bitstring without losing information, so that it can be later retrieved. Then *Shannon's noiseless coding theorem* affirms that asymptotically—i.e. for diverging n—the amount of bits needed per source symbol is given by $H(X)$.

More formally, if $\ell_{compr}^\varepsilon(X)$ is the minimum amount of bits needed to compress X without losing information, except for probability ε, then the *compression rate* of the example above is given by:

$$r_{compr}(X) := \lim_{\varepsilon \to 0} \lim_{n \to \infty} \frac{\ell_{compr}^\varepsilon(X_1 X_2 \cdots X_n)}{n} = H(X). \tag{2.53}$$

Similarly, we consider the quantum i.i.d. source defined by the state ρ, with spectral decomposition $\rho = \sum_i \lambda_i |\psi_i\rangle\langle\psi_i|$. In other words, the source emits a sequence of quantum states drawn from $\{|\psi_i\rangle\}$, according to the distribution $\{p_i\}$. For *Schumacher's noiseless coding theorem*, the fraction of qubits needed to reliably encode and decode each state in the sequence is given by the von Neumann entropy $H(\rho)$.

However, if one removes the asymptotic or the i.i.d. assumption, the Shannon and von Neumann entropies no longer describe operational quantities. Consider for instance a single realization (non-asymptotic regime) of a random variable X representing an n-bit string. With probability $1/2$ the string is composed of all zeroes and with probability $1/2$ the string is random and uniformly distributed. The probability distribution governing the outcomes of X is thus $\{\frac{1}{2}, \frac{1}{2^{n+1}}, \ldots, \frac{1}{2^{n+1}}\}$. The Shannon entropy of X is given by: $H(X) = \frac{n}{2} + 1$. Conversely the minimum compression length of X, for a single realization of X, is $\ell_{\text{compr}}^{\varepsilon}(X) \approx n$ for a sufficiently small ε [12]. This example shows that the Shannon entropy may arbitrarily deviate from the operational quantity it represents in non-asymptotic or non-i.i.d. scenarios.

2.7 Min- and Max-Entropy

In this Section we introduce two additional entropy measures that play a fundamental role in the security of quantum cryptographic schemes. Moreover, the smoothed versions of such entropies can be regarded as generalizations of the Shannon and von Neumann entropies to non-i.i.d. and non-asymptotic scenarios.

Definition 2.8 (*Min-entropy* [12, 13]) Let ρ_{AB} be a bipartite density operator. The min-entropy of A conditioned on B of the state ρ_{AB} is defined as:

$$H_{\min}(A|B)_{\rho} = -\log \min_{\sigma_B}\{\operatorname{Tr}(\sigma_B) \ : \ \sigma_B \geq 0, \ (\mathbb{1}_A \otimes \sigma_B) - \rho_{AB} \geq 0\}. \qquad (2.54)$$

Definition 2.9 (*Max-entropy* [12, 13]) Let ρ_{AB} be a bipartite density operator and let ρ_{ABC} be a purification of ρ_{AB}. The max-entropy of A conditioned on B of the state ρ_{AB} is defined as:

$$H_{\max}(A|B)_{\rho} = -H_{\min}(A|C)_{\rho}. \qquad (2.55)$$

When the system B is trivial, i.e. one-dimensional, the conditioning is omitted and the entropies are represented by $H_{\min}(A)$ and $H_{\max}(A)$. For instance, the min-entropy of a quantum state ρ_A reduces to: $H(A)_{\rho} = -\log \lambda_{\max}(\rho_A)$, where $\lambda_{\max}(\rho_A)$ is the largest eigenvalue of ρ_A, i.e the largest probability value in the probability distribution defined by the eigenvalues of ρ_A. Min- and max-entropy are also defined on a probability distribution P_X by evaluating them on the state: $\rho_X = \sum_x P_X(x)|x\rangle\langle x|$, where $\{|x\rangle\}$ is an orthonormal basis.

In general, the entropies are related as follows to the von Neumann entropy of a bipartite density operator ρ_{AB} [14]:

$$H_{\min}(A|B) \leq H(A|B) \leq H_{\max}(A|B). \qquad (2.56)$$

2.7.1 Operational Meaning of Min-Entropy

The following operational interpretation of the min-entropy suggests the importance of this entropy measure for quantum cryptography.

Consider the following adversarial scenario. An honest party, Alice, is in possess of a random key K that she would like to keep secret. An adversary, Eve, wants to learn Alice's key. In order to do so, Eve holds a quantum system E whose state is correlated with the value of Alice's key K. This scenario is represented by the c.q. state:

$$\rho_{KE} = \sum_{k \in \mathcal{K}} p_k \, |k\rangle \langle k| \otimes \rho_E^k, \tag{2.57}$$

where $\{|k\rangle\}$ is a orthonormal set of vectors representing Alice's possible keys[2] and $\{p_k\}$ is their probability distribution. Eve attempts to learn the value k of the key by performing a suitable measurement on her system in state ρ_E^k. Let $p_{\text{guess}}(K|E)$ be the probability that Eve correctly guesses K when using an optimal measurement strategy, i.e.:

$$p_{\text{guess}}(K|E) = \max_{\{E_k\}} \sum_{k \in \mathcal{K}} p_k \, \mathrm{Tr}[E_k \rho_E^k], \tag{2.58}$$

where $\{E_k\}$ are POVM elements of a generic quantum measurement on system E. Then, the min-entropy of the state (2.57) is related to Eve's *guessing probability* $p_{\text{guess}}(K|E)$ by [12]:

$$H_{\min}(K|E)_\rho = -\log p_{\text{guess}}(K|E). \tag{2.59}$$

Note that if one removes the conditioning on E, the min-entropy reads $H_{\min}(K) = -\log \max_k p_k$ and can still be linked to the optimal guessing probability. Indeed, in absence of side information on K, the best guess one can make on its value is the key k occurring with the highest probability, that is $\max_k p_k$.

> How could Alice use the information on Eve's guessing probability, i.e. the min-entropy $H_{\min}(K|E)$, in order to increase the secrecy of her key?

We are going to provide the complete answer in Sect. 2.10 with *privacy amplification*. Nevertheless, here we provide an intuitive and informal answer.

Assume that the adversarial scenario described above is replicated many times, where each instance is described by the same c.q. state (2.57). Then, in some cases Eve correctly guesses Alice's key K, and in others she does not. In this setting, the guessing probability $p_{\text{guess}}(K|E)$ may be interpreted as the frequency of Eve's

---

[2]Alice's keys are encoded in orthogonal states since they are classical bitstrings and therefore perfectly distinguishable (see Sect. 2.11).

correct guesses, i.e. the number of keys n_g guessed by Eve over the total number of keys n. Then the min-entropy is just given by: $H_{\min}(K|E)_\rho = \log n - \log n_g$. That is, it represents the difference between the total number of bits composing each of the n possible keys and the number of bits encoding the guessed keys. Therefore the min-entropy quantifies, in bits, the average amount of information of each key which is not learned by Eve and is thus secret. With this information, Alice applies a privacy amplification procedure and distils a "completely" secret key from her initial key K (see Sect. 2.10).

2.8   Smooth Min- and Max-Entropy

In order to account for an error probability ε in the information-processing tasks related to min- and max-entropy, we introduce the ε-smooth versions of the entropies. The ε-smooth min- and max-entropy of a quantum state ρ_{AB} can be interpreted as generalizations of the Shannon/von Neumann entropy (Sect. 2.6). They are obtained by optimizing the corresponding non-smooth entropies over a ball of states which are close to ρ_{AB} according to the notion of purified distance $P(\rho, \tau)$ (c.f. Sect. 2.11 in the Appendix of this Chapter).

Definition 2.10 (*Smooth entropies* [12, 13]) Let ρ_{AB} be a bipartite density operator. The ε-smooth min- and max-entropy of A conditioned on B of the state ρ_{AB} are given by:

$$H_{\min}^{\varepsilon}(A|B)_\rho = \max_{\sigma \in \mathcal{B}^\varepsilon(\rho_{AB})} H_{\min}(A|B)_\sigma \qquad (2.60)$$

$$H_{\max}^{\varepsilon}(A|B)_\rho = \min_{\sigma \in \mathcal{B}^\varepsilon(\rho_{AB})} H_{\max}(A|B)_\sigma, \qquad (2.61)$$

where $\mathcal{B}^\varepsilon(\rho_{AB})$ is a ball of ε-close states centred in ρ_{AB}:

$$\mathcal{B}^\varepsilon(\rho_{AB}) = \{\tau_{AB} \geq 0 : \mathrm{Tr}(\tau_{AB}) \leq 1, \ P(\rho_{AB}, \tau_{AB}) \leq \varepsilon\}. \qquad (2.62)$$

The *asymptotic equipartition property* (AEP) links the smooth entropies to the Shannon/von Neumann entropy [14]:

$$H(A|B)_\rho = \lim_{\varepsilon \to 0} \lim_{n \to \infty} \frac{1}{n} H_{\min}^{\varepsilon}(A^n|B^n)_{\rho^{\otimes n}} \qquad (2.63)$$

$$H(A|B)_\rho = \lim_{\varepsilon \to 0} \lim_{n \to \infty} \frac{1}{n} H_{\max}^{\varepsilon}(A^n|B^n)_{\rho^{\otimes n}}, \qquad (2.64)$$

where the smooth entropies are evaluated on the i.i.d. state $\rho^{\otimes n}$. Another important property of the smooth entropies is the *data-processing inequality* [13]:

$$H_{\min}^{\varepsilon}(A|B)_{\rho} \leq H_{\min}^{\varepsilon}(A|B')_{(\mathbb{1}_A \otimes \mathcal{E}_B)\rho} \qquad (2.65)$$

$$H_{\max}^{\varepsilon}(A|B)_{\rho} \leq H_{\max}^{\varepsilon}(A|B')_{(\mathbb{1}_A \otimes \mathcal{E}_B)\rho}. \qquad (2.66)$$

The data-processing inequality basically states that if we process the quantum side information B through a CP trace-preserving map $\mathcal{E}$, we always increase our uncertainty on A.

The smooth min- and max-entropy are well suited to characterize operational quantities in realistic scenarios (e.g.. finite resources and errors), which often appear in quantum cryptographic schemes. In the following two Sections we provide some examples.

2.9 Data Compression and Error Correction

Recall that $\ell_{\mathrm{compr}}^{\varepsilon}(X)$ is the minimum amount of bits encoding a single realization of the random variable X, from which the value of X can be recovered with probability at least $1 - \varepsilon$. This quantity is essentially equal to the ε-smooth max-entropy of the distribution P_X [12]:

$$\ell_{\mathrm{compr}}^{\varepsilon}(X) = H_{\max}^{\varepsilon'}(X) + O(\log 1/\varepsilon), \qquad (2.67)$$

for some $\varepsilon' \in [\frac{\varepsilon}{2}, 2\varepsilon]$. This result generalizes Shannon's noiseless coding theorem (2.53) to a scenario where the number of realizations of X is finite. Shannon's theorem is recovered by employing (2.67) in (2.53) and by using the AEP (2.64).

The result in (2.67) can be applied to the cryptographic scenario where two parties, Alice and Bob, establish a shared secret key (bitstring) over a noisy channel. Due to the noise, Bob has only a probability distribution $P_{X|Y}$ of the possible keys X held by Alice, conditioned on his noisy side information Y. By performing a classical error correction (EC) procedure, Alice and Bob aim at sharing the same secret key. For instance, Alice sends to Bob the minimal amount of information $\ell_{\mathrm{compr}}^{\varepsilon}(X|Y)$ that allows him to correctly guess her key, except for probability ε. This information is equal to the smallest reliable data compression of X, when Y is known, i.e. $\ell_{\mathrm{compr}}^{\varepsilon}(X|Y) \approx H_{\max}^{\varepsilon'}(X|Y)$.

2.10 Privacy Amplification

Consider the same adversarial scenario described in Sect. 2.7.1. Alice holds a random key K correlated with a quantum system E held by the eavesdropper Eve, as described by Eq. (2.57). Eve attempts to learn Alice's key by properly measuring her quantum system E.

The goal of privacy amplification (PA) is to extract a *secret key S* by applying a random function F on K: $S = F(K)$. We define a key to be secret if it is distributed *uniformly* and is *independent* of Eve's quantum side information E.

Ideally, PA is successful if the quantum state describing Alice's key S and Eve's side information EF (we allow Eve to know which function Alice applied on her key K) is given by:

$$\rho_{SEF} = \omega_S \otimes \rho_{EF}, \tag{2.68}$$

where the state of the secret key S is uniformly distributed over the possible final keys $s \in S$ of Alice:

$$\omega_S = \frac{1}{|S|} \sum_{s \in S} |s\rangle\langle s|. \tag{2.69}$$

Indeed, in ρ_{SEF} Eve's system is no longer correlated with the state $|s\rangle$ of Alice's key after PA, hence the key S is secret.

In reality, we relax the claim on the success of PA by allowing the final state ρ_{SEF} to be "almost" indistinguishable from the ideal scenario given by the r.h.s. of (2.68). Specifically, we require ρ_{SEF} to be ε-indistinguishable (see Sect. 2.11) from $\omega_S \otimes \rho_{EF}$. This translates to an upper bound on the trace distance between the two states (c.f. Sect. 2.11):

$$T(\rho_{SEF}, \omega_S \otimes \rho_{EF}) \leq \varepsilon. \tag{2.70}$$

In order for Alice to achieve the PA goal stated in (2.70), she applies to her key K a hash function f from a two-universal family: $f \in_R \mathcal{F}$, where "$\in_R$" indicates that the function is randomly picked, with probability $1/|\mathcal{F}|$, from the family $\mathcal{F}$.

Definition 2.11 (*Two-universal family* [4, 15]) A family of hash functions $\mathcal{F} = \{f \text{ s.t. } f : \mathcal{K} \to S\}$ is two-universal if, for every pair of keys $k_1, k_2 \in \mathcal{K}$ such that $k_1 \neq k_2$, it holds[3]:

$$\Pr_{f \in_R \mathcal{F}} [f(k_1) = f(k_2)] \leq \frac{1}{|S|}. \tag{2.71}$$

The state ρ_{SEF} of Alice's final key and Eve's information after PA reads:

$$\rho_{SEF} = \sum_{f \in \mathcal{F}} \frac{1}{|\mathcal{F}|} \rho_{f(K)E} \otimes |f\rangle\langle f|_F, \tag{2.72}$$

where the state $\rho_{f(K)E}$ is obtained from ρ_{KE} in (2.57) by specifically applying the hash function f to the key K, and is given by:

---

[3] Note that the number of functions in a two-universal family, $|\mathcal{F}|$, must be in general larger than the number of keys they can generate, $|S|$. Indeed: $\Pr_{f \in_R \mathcal{F}} [f(k_1) = f(k_2)] = \frac{|\{f \in \mathcal{F} : f(k_1) = f(k_2)\}|}{|\mathcal{F}|} \leq \frac{1}{|S|}$, which implies that $|\mathcal{F}| \geq |S|$.

$$\rho_{f(K)E} = \sum_{s \in S} |s\rangle\langle s| \otimes \rho_E^s \quad , \quad \rho_E^s = \sum_{k \in f^{-1}(s)} p_k \rho_E^k. \qquad (2.73)$$

Let $\ell = \log|S|$ be the number of bits of the secret key S generated by Alice when applying PA on the state ρ_{KE}. Then the *Quantum Leftover Hash Lemma* [16] provides an upper bound on the trace distance between the real state ρ_{SEF} and the ideal state $\omega_S \otimes \rho_{EF}$ as a function of ℓ and of the smooth min-entropy of the original state[4] ρ_{KE}.

Lemma 2.1 (Quantum Leftover Hash Lemma [16]) *Let ρ_{KE} be a c.q. state of the form (2.57). Let ρ_{SEF} be the state (2.72) obtained from ρ_{KE} by applying a random two-universal hash function on K. Then for every ε' it holds:*

$$T(\rho_{SEF}, \omega_S \otimes \rho_{EF}) \leq 2\varepsilon' + \frac{1}{2}\sqrt{2^{\ell - H_{\min}^{\varepsilon'}(K|E)}}. \qquad (2.74)$$

By combining the above Lemma with the requirement that the PA outcome is ε-indistinguishable from an ideal one (2.70), one obtains an upper bound on the length ℓ of the secret key S in terms of the smooth min-entropy of the original state ρ_{KE}:

$$\ell \leq H_{\min}^{\varepsilon'}(K|E) + 2 - 2\log\frac{1}{\varepsilon - 2\varepsilon'}, \qquad (2.75)$$

The bound (2.75) can be optimized over ε', with $\varepsilon' \in [0, \varepsilon/2)$.

As anticipated, the smooth min entropy possesses an important operational meaning in the field of quantum cryptography (2.75).

> Let ρ_{KE} be a c.q. state describing an insecure key K and the eavesdropper's quantum side information E. The smooth min-entropy $H_{\min}^{\varepsilon'}(K|E)$ quantifies the bit-length of the ε-indistinguishable secret key extracted from K by PA, up to corrections of order $O(\log 1/\varepsilon)$.

Both EC and PA are fundamental tasks in any quantum key distribution (QKD) protocol. Indeed, as we shall see in the next Chapter, the final secret key length of a generic QKD protocol is determined by a combination of smooth min- and max-entropy.

---

[4]We remark that the result in [16] is actually weaker and provides an upper bound on $\min_{\sigma_{EF}} T(\rho_{SEF}, \omega_S \otimes \sigma_{EF})$. Nevertheless, the result stated in Lemma 2.1 is also valid and can be proven with analogous steps to those of [16].

Appendix

In this Appendix we define two notions of distance between quantum states that are often used in quantum cryptography. We also provide a definition of ε-indistinguishability of two quantum states.

2.11 Distances and Distinguishability Between Quantum States

We first define a commonly used norm for linear operators, the trace norm.

Definition 2.12 (*Trace norm* [2]) The trace norm of an operator O is defined as the sum of its singular values, or equivalently as:

$$\|O\| = \mathrm{Tr}\left[\sqrt{OO^\dagger}\right]. \tag{2.76}$$

Definition 2.13 (*Trace distance* [2]) The trace distance between two density operators ρ and τ is proportional to the trace norm of their difference:

$$T(\rho, \tau) = \frac{1}{2}\|\rho - \tau\| = \frac{1}{2}\mathrm{Tr}\left[\sqrt{(\rho - \tau)^2}\right] = \frac{1}{2}\sum_i |\lambda_i|, \tag{2.77}$$

where λ_i are the eigenvalues of the Hermitian operator $\rho - \tau$.

The trace distance has an operational interpretation linked to the distinguishability of quantum states. Consider Alice preparing a system in either the state ρ or the state τ, each with probability $1/2$. Bob receives the system and is asked to discriminate between the two states by measuring the system with a binary POVM $\{P_0, \mathbb{1} - P_0\}$. Bob arbitrarily assigns the outcome P_0 to the detection of state ρ, and outcome $\mathbb{1} - P_0$ to the detection of state τ. The conditional probability of Bob obtaining outcome P_0, given that Alice prepared state ρ, is: $p(0|\rho) = \mathrm{Tr}[P_0\rho]$. Then, the probability that Bob correctly guesses the state prepared by Alice is given by:

$$\begin{aligned}
p_{\text{guess}}(\rho, \tau) &= \frac{1}{2}p(0|\rho) + \frac{1}{2}p(1|\tau) \\
&= \frac{1}{2}\left(\mathrm{Tr}[P_0\rho] + \mathrm{Tr}[(\mathbb{1} - P_0)\tau]\right) \\
&= \frac{1}{2}\left(1 + \mathrm{Tr}[P_0(\rho - \tau)]\right).
\end{aligned} \tag{2.78}$$

If Bob employs the optimal measurement strategy, i.e. if $\mathrm{Tr}[P_0(\rho - \tau)]$ is maximized over the possible POVMs of the form $\{P_0, \mathbb{1} - P_0\}$, one can relate the optimal guessing probability of Bob to the trace distance between the two states prepared by Alice [2]:

$$p_{\text{guess}}(\rho, \tau) = \frac{1}{2}\left(1 + T(\rho, \tau)\right). \qquad (2.79)$$

If $T(\rho, \tau) = 0$ the two states are the same state and the corresponding optimal guessing probability reads $p_{\text{guess}}(\rho, \tau) = 1/2$. This is expected, since Bob cannot perform a better guess of the received state than a random guess (e.g., a coin flip), when the two states are identical. We thus call the two states *indistinguishable*. In general, Eq. 2.79 shows that the trace distance $T(\rho, \tau)$ between two states coincides with the *distinguishing advantage* of a distinguisher (in our case Bob) attempting to distinguish them [17].

We formalize the situation where two states are "almost" indistinguishable with the following definition.

Definition 2.14 (*States indistinguishability*) The states ρ and τ are ε-indistinguishable if their trace distance is upper bounded by:

$$T(\rho, \tau) \leq \varepsilon \qquad (2.80)$$

i.e. if the states are indistinguishable, except for a probability at most $\varepsilon/2$.

We derive a closed expression for the trace distance between two pure states $|\psi\rangle$ and $|\phi\rangle$:

$$T(\psi, \phi) = \frac{1}{2}\,\||\psi\rangle\langle\psi| - |\phi\rangle\langle\phi|\| = \frac{1}{2}\sum_i |\lambda_i|, \qquad (2.81)$$

by computing the eigenvalues λ_i of the operator $O := |\psi\rangle\langle\psi| - |\phi\rangle\langle\phi|$. The eigenvalue equation of O reads:

$$O|\lambda_i\rangle = |\psi\rangle\langle\psi|\lambda_i\rangle - |\phi\rangle\langle\phi|\lambda_i\rangle = \lambda_i|\lambda_i\rangle. \qquad (2.82)$$

We now take the inner product between each of the two pure states and both sides in (2.82):

$$\begin{cases} \langle\psi|\lambda_i\rangle - \langle\psi|\phi\rangle\langle\phi|\lambda_i\rangle = \lambda_i\langle\psi|\lambda_i\rangle \\ \langle\phi|\psi\rangle\langle\psi|\lambda_i\rangle - \langle\phi|\lambda_i\rangle = \lambda_i\langle\phi|\lambda_i\rangle \end{cases} \qquad (2.83)$$

The above linear system in the variables $\langle\psi|\lambda_i\rangle$ and $\langle\phi|\lambda_i\rangle$ can be recast as:

$$\begin{cases} (1 - \lambda_i)\langle\psi|\lambda_i\rangle - \langle\psi|\phi\rangle\langle\phi|\lambda_i\rangle = 0 \\ \langle\phi|\psi\rangle\langle\psi|\lambda_i\rangle - (1 + \lambda_i)\langle\phi|\lambda_i\rangle = 0 \end{cases} \qquad (2.84)$$

The above homogeneous system is solvable if the two equations are linearly dependent:

$$\det \begin{bmatrix} 1 - \lambda_i & -\langle \psi | \phi \rangle \\ \langle \psi | \phi \rangle & -(1 + \lambda_i) \end{bmatrix} = 0, \tag{2.85}$$

which is verified for $\lambda_i = \pm\sqrt{1 - |\langle \psi | \phi \rangle|^2}$. By substituting the expression for λ_i in (2.81) we get the following result.

The trace distance between two pure states $|\psi\rangle$ and $|\phi\rangle$ is a function of their overlap (inner product):

$$T(\psi, \phi) = \sqrt{1 - |\langle \psi | \phi \rangle|^2}. \tag{2.86}$$

Equation (2.79) combined with (2.86) indicate that the higher the overlap between two pure states, the less distinguishable they are, as expected.

The notion of distance between quantum states that is commonly used in defining the ε-smooth min- and max-entropy is the *purified distance*.

Definition 2.15 (*Purified distance* [16]) The purified distance between two positive operators ρ and τ is given by:

$$P(\rho, \tau) = \sqrt{1 - \overline{F}(\rho, \tau)^2}, \tag{2.87}$$

where $\overline{F}(\rho, \tau)$ is the generalized fidelity:

$$\overline{F}(\rho, \tau) = \left\| \sqrt{\tau}\sqrt{\rho} \right\| + \sqrt{(1 - \operatorname{Tr} \rho)(1 - \operatorname{Tr} \sigma)}. \tag{2.88}$$

An important property of the purified distance, which is not satisfied by the trace distance, is that if two states are separated by a distance $P(\rho, \tau)$, there exist purifications of ρ and τ with the same purified distance.

References

1. Kaye, P., Laflamme, R., & Mosca, M. (2007). *An introduction to quantum computing*. Oxford University Press.
2. Nielsen, M. A., & Chuang, I. L. (2010). *Quantum computation and quantum information* (10th Anniversary ed.). Cambridge University Press.
3. Rossetti, C. (2011). *Rudimenti di meccanica quantistica*. Levrotto & Bella.

4. Renner, R. (2008). Security of quantum key distribution. *International Journal of Quantum Information, 06*(01), 1–127.
5. Yanai, H., Takeuchi, K., & Takane, Y. (2011). *Projection matrices*, pp. 25–54. New York, NY: Springer.
6. Peres, A. (2006). *Quantum theory: Concepts and methods*. Dordrecht: Springer.
7. Paris, M. G. A. (2012). The modern tools of quantum mechanics. *The European Physical Journal Special Topics, 203*(1), 61–86.
8. Pauli, W. (1927). Zur quantenmechanik des magnetischen elektrons. *Zeitschrift für Physik, 43*(9), 601–623.
9. Gühne, O., & Tóth, G. (2009). Entanglement detection. *Physics Reports, 474*(1), 1–75.
10. Horodecki, R., Horodecki, P., Horodecki, M., & Horodecki, K. (2009). Quantum entanglement. *Reviews of Modern Physics, 81*, 865–942.
11. Greenberger, D. M., Horne, M. A., & Zeilinger, A. (1989). *Bell's theorem, quantum theory, and conceptions of the Universe*, vol. 37. Springer Science & Business Media B.V.
12. Konig, R., Renner, R., & Schaffner, C. (2009). The operational meaning of min- and max-entropy. *IEEE Transactions on Information Theory, 55*(9), 4337–4347.
13. Tomamichel, M. (2016). *Quantum information processing with finite resources*. Springer Briefs in Mathematical Physics.
14. Tomamichel, M., Colbeck, R., & Renner, R. (2009). A fully quantum asymptotic equipartition property. *IEEE Transactions on Information Theory, 55*(12), 5840–5847.
15. Carter, J., & Wegman, M. N. (1979). Universal classes of hash functions. *Journal of Computer and System Sciences, 18*(2), 143–154.
16. Tomamichel, M., Schaffner, C., Smith, A., & Renner, R. (2011). Leftover hashing against quantum side information. *IEEE Transactions on Information Theory, 57*(8), 5524–5535.
17. Portmann, C. and Renner, R. (2014). Cryptographic security of quantum key distribution. arXiv:quant-ph/1409.3525.

Chapter 3
Introducing Quantum Key Distribution

As the need for unbreakable encryption looms in networks around the world, quantum cryptography is the solution that will safeguard and future-proof sensitive information. Commercial QKD company

Abstract Quantum key distribution (QKD) is arguably the most developed task of quantum cryptography, both from a theoretical and experimental point of view. In this Chapter we first present some of the security principles of QKD in Sect. 3.1. We then describe the BB84 protocol as an example of QKD protocol and compute its key rate (Sect. 3.2). Section 3.3 is bit more technical, here we define and prove the security of a generic QKD protocol in the finite-key scenario. We conclude the Chapter by providing an incomplete overview of the most recent state-of-the-art QKD experiments (Sect. 3.4). In the Appendix (Sects. 3.5 and 3.6) we report the details of some proofs of the main text.

The security of classical cryptographic schemes relies on assumptions on the adversary's computational capabilities and on the fact that certain mathematical problems are considered "hard" to solve. This makes classical cryptography vulnerable to retroactive attacks. That is, an adversary could store the encrypted data while it is communicated and wait to have sufficient computational power, or smarter algorithms, in order to decrypt it. Conversely, the security of quantum cryptography relies on intrinsic principles of nature, as described by quantum mechanics. Therefore, assuming that quantum mechanics is correct, the security offered by quantum cryptography is everlasting, in the sense that it is independent of future theoretical or experimental advances of the adversary.

F. Grasselli, *Quantum Cryptography*, Quantum Science and Technology,
https://doi.org/10.1007/978-3-030-64360-7_3

3.1 The Origins of Security

QKD is a specific task of quantum cryptography where two honest parties, traditionally called Alice and Bob, establish a shared secret key when connected by an insecure quantum channel and an authenticated public classical channel. The combination of QKD with the Vernam cipher [1, 2], also called one-time pad, allows for ever-lasting secure communication.

Indeed, suppose that Alice wants to send a secret message $\mathbf{m}$, composed of n bits, to Bob. According to the Vernam cipher, Alice encrypts her message by adding it modulo two[1] with a n-bit key $\mathbf{k}$ she shares with Bob, thanks to a prior execution of a QKD protocol: $\mathbf{m}_e = \mathbf{m} \oplus \mathbf{k}$. She then sends the encrypted message $\mathbf{m}_e$ to Bob, who decrypts it by again adding the encryption key: $\mathbf{m}_e \oplus \mathbf{k} = \mathbf{m} \oplus \mathbf{k} \oplus \mathbf{k} = \mathbf{m}$. The Vernam cipher is provably secure as long as the number of key bits matches the number of message bits, and the key (or parts of it) is not reused [3]. The security of the communication thus depends on the security of the QKD protocol.

Many QKD protocols are based on the transmission of quantum states from Alice to Bob, through the quantum channel. The crucial fact which makes QKD secure is that a potential eavesdropper, Eve, cannot gain any information from the transmitted states without disturbing them.

For instance, an obvious attack by Eve would be to create perfect copies of the transmitted states before they reach Bob, as in classical wiretapping. However, quantum mechanics prevents this, as shown by the *no-cloning theorem*.

Theorem 3.1 (no-cloning [4]) *It is not possible to perfectly clone an unknown quantum state.*

Proof Suppose by contradiction that we have a cloning machine and that we apply it on two distinct quantum states $|\psi\rangle \neq |\phi\rangle$ which are also non-orthogonal $\langle\phi|\psi\rangle \neq 0$. The action of the cloning machine is represented by a unitary operation U, which copies the input state on some auxiliary system initially in a normalized state $|s\rangle$:

$$U(|\psi\rangle \otimes |s\rangle) = |\psi\rangle \otimes |\psi\rangle \tag{3.1}$$

$$U(|\phi\rangle \otimes |s\rangle) = |\phi\rangle \otimes |\phi\rangle. \tag{3.2}$$

By taking the inner product of equations (3.1) and (3.2) we obtain:

$$\langle\phi|\psi\rangle = (\langle\phi|\psi\rangle)^2, \tag{3.3}$$

which is only true when the states $|\psi\rangle$ and $|\phi\rangle$ are either the same state or are orthogonal, thus a general cloning machine is not possible.   □

We remark that this theorem does not contradict our common sense that classical information can be copied, since the latter is always stored in physical systems (e.g., a

---

[1]In this case the "$\oplus$" symbol indicates the XOR operation on bits or bitstrings.

piece of written paper) described by orthogonal quantum states. As the proof showed, quantum mechanics does not prevent to build a machine which clones orthogonal quantum states.

Considered that Eve cannot copy non-orthogonal transmitted states, at least she would like to be able to partially distinguish them, without being noticed. However, this is also forbidden by quantum mechanics.

Proposition 3.1 (Information gain entails disturbance [3]) *In the attempt to distinguish non-orthogonal quantum states in a quantum signal, any information gain is accompanied by a disturbance of the signal.*

Proof Let $|\psi\rangle$ and $|\phi\rangle$ be two non-orthogonal quantum states in the quantum signal sent by Alice to Bob. Eve's action on the signal is represented by a generic quantum operation, which can be viewed as a unitary acting on a larger Hilbert space (c.f. Sect. 2.5). In particular, the unitary acts on the state $|\psi\rangle$ (or $|\phi\rangle$) and on an ancilla $|u\rangle$. We assume that Eve's action leaves the signal states unchanged:

$$U(|\psi\rangle \otimes |u\rangle) = |\psi\rangle \otimes |v\rangle \tag{3.4}$$
$$U(|\phi\rangle \otimes |u\rangle) = |\phi\rangle \otimes |v'\rangle. \tag{3.5}$$

Eve would like $|v\rangle$ and $|v'\rangle$ to be different states, so she could partially distinguish the corresponding signal states. However, by computing the inner product of equations (3.4) and (3.5) we obtain that:

$$\langle\phi|\psi\rangle = \langle\phi|\psi\rangle\langle v'|v\rangle, \tag{3.6}$$

implying that $|v\rangle = |v'\rangle$. Thus, distinguishing two non-orthogonal states implies the disturbance of at least one of them. □

The above results suggest how quantum mechanical properties can be exploited in a key distribution scheme. Alice can encode the key bits in non-orthogonal quantum states and send them to Bob. By checking the disturbance of the signal, the parties can quantitatively upper bound Eve's knowledge on the exchanged key.

3.2 The BB84 Protocol

The BB84 protocol [5], named after its inventors Bennett and Brassard, is commonly considered to be the first ever QKD protocol, but it is also the simplest and variations of it are investigated and implemented even today. For the protocol's description, we follow the references [6, 7].

Suppose Alice possesses a source of single photons, whose spectral properties are well defined so that the only remaining degree of freedom is the photon's polarization. Alice and Bob align their polarizers and agree to employ two polarization bases, one defined by the horizontal/vertical directions $(0°/90°)$ and the other defined by the diagonal/antidiagonal $(+45°/-45°)$ directions. The polarization state of a photon is thus represented by a qubit in $\mathcal{H}_2$. We associate the eigenbasis $\{|0\rangle, |1\rangle\}$ of Pauli operator Z to the horizontal/vertical basis and the eigenbasis $\{|+\rangle, |-\rangle\}$ of X to the diagonal/antidiagonal basis, where $|\pm\rangle = (|0\rangle \pm |1\rangle)/\sqrt{2}$. The BB84 protocol comprises the following steps:

1. Alice sends to Bob a sequence of M photons randomly prepared in one of the four states $|0\rangle, |1\rangle, |+\rangle$ and $|-\rangle$, via the quantum channel. The parties identify the bit value 0 (1) with the non-orthogonal states $|0\rangle$ and $|+\rangle$ ($|1\rangle$ and $|-\rangle$). The non-orthogonality condition ensures that any tampering with the quantum channel by Eve, in order to gain information on the transmitted key, leads to a disturbance of the signal and can be later detected by the parties.
2. Upon receiving a photon, Bob measures randomly in either the Z or the X basis. If Bob measures in the same basis Alice used to prepare the photon, he learns the bit she encoded on that photon, provided that the signal has not been altered. If instead Bob measures in the complementary basis, he obtains a random bit since the two bases are mutually unbiased (c.f. Sect. 2.1).
3. **Sifting** Once the quantum communication is over, Alice and Bob publicly compare the bases they used on each photon and discard the bits corresponding to unmatching bases. This process leaves Alice and Bob with strings of approximately $M/2$ bits. In absence of errors due to noise or eavesdropping, the bitstrings of Alice and Bob would coincide.
4. **Parameter estimation (PE)** Alice and Bob reveal a random sample of their bits in order to estimate the error rate in the quantum channel and thus the information gained by Eve.[2] In particular, the parties estimate the quantum bit error rate (QBER) in the Z (X) basis, i.e. the fraction E_Z (E_X) of bits generated by measuring in the Z (X) basis that disagree. The computed QBERs are the input parameters of the following steps. The parties are now left with two partially-correlated and partially-secret bitstrings, called the *raw keys*. We denote a generic raw key bit of Alice (Bob) by the random variable R_A (R_B).
5. **Error correction (EC)** Alice and Bob run a *one-way error correction* algorithm to correct Bob's raw key to match Alice's. Alice sends the required information over a classical public channel to Bob. Other EC schemes are possible.
6. **Privacy amplification (PA)** The parties remove the information that Eve gained on their error-corrected keys by compressing them to a shorter secret key via a randomness extractor (e.g., two-universal hashing, Sect. 2.10).

The figure of merit of every QKD protocol is the *secret key rate*, i.e. the fraction of secure key bits produced per protocol round.[3] A round is commonly regarded as the

---

[2]For security reasons one needs to consider the worst-case scenario, i.e. that all the noise in the channel is due to Eve.

[3]In experiments, the secret key rate is often given in terms of secret key bits per second. This is obtained by multiplying the secret key rate defined here by the repetition rate of the protocol, i.e. the number of protocol rounds per second.

transmission of a quantum state through the quantum channel. The secret key rate generally depends on the total number of rounds M performed.

In the following we compute the secret key rate of the BB84 protocol in the *asymptotic scenario* of infinitely many rounds: $M \to \infty$. This is, of course, an unrealistic assumption, but it greatly simplifies the math. Moreover, the asymptotic key rate is often used as a benchmark for the performance of a newly-developed QKD protocol.

The protocol above is presented in *prepare-and-measure* form, since one party prepares and sends quantum states while the other measures them. This is typically what happens in real-life implementations of many QKD protocols. However, when proving the security of a QKD protocol or computing its key rate, an equivalent *entanglement-based* description is much more convenient.

Ideally, in every round of the entanglement-based BB84 protocol the two-qubit Bell state

$$|\Phi^+\rangle_{AB} = \frac{|00\rangle + |11\rangle}{\sqrt{2}} = \frac{|++\rangle + |--\rangle}{\sqrt{2}} \tag{3.7}$$

is generated, and the two qubits are distributed to Alice and Bob through the quantum channel. Alice and Bob then measure the received qubit in either the Z or X basis, obtaining the same outcome if they chose the same basis. This scenario is equivalent to the prepare-and-measure one since the state Bob receives, conditioned on Alice measuring e.g., X and obtaining outcome x, is exactly $|x\rangle$ where $x \in \{+, -\}$.

However, in reality Eve could be in total in control of the quantum channel, distributing a mixed state ρ_{AB} to the parties in every protocol round. We assign to Eve all the information that can be correlated with the mixed state ρ_{AB} by assuming that she holds the purifying system E (recall Sect. 2.4.1). That is, the state on A, B and E is pure: $|\phi_{ABE}\rangle$. In this scenario we say that Eve performs a *collective attack* and the quantum state representing Alice and Bob's qubits in the M protocol rounds is the i.i.d. state $\rho_{AB}^{\otimes M}$. The parties detect the presence of Eve from the errors (E_Z and E_X) in the outcomes generated by measuring ρ_{AB} at every round.

There is even a more general scenario where Eve directly distributes the state ρ_{AB}^M describing all the M qubit pairs to be measured, of which she holds the purifying system E, i.e. ρ_{ABE}^M is pure. In this case Eve is performing a *coherent attack*, which is generally more powerful than collective attacks since the states shared by Alice and Bob in each round can be correlated with past and future rounds—formally, it holds: $\rho_{AB}^M \neq \rho_{AB}^{\otimes M}$.

We address the case of coherent attacks in the next Section, where we investigate the security of QKD when the number of protocol rounds is finite. Conversely, in the asymptotic regime discussed here, the two attacks are proven to be equivalent (c.f. Sect. 3.3.3), hence we restrict to collective attacks and focus on one specific protocol round.

3.2.1 Secret Key Rate

The asymptotic secret key rate of any QKD protocol with one-way EC is given by the Devetak-Winter rate [8]:

$$r_{\text{DW}} = H(R_A : R_B) - H(R_A : E), \tag{3.8}$$

which can be recast in the more familiar form [9–11]:

$$r = H(R_A|E) - H(R_A|R_B), \tag{3.9}$$

by using the definition of mutual information (c.f. Sect. 2.6). Recall that R_A and R_B are the random variables representing Alice's and Bob's raw key bit.

An intuitive explanation of the key rate expression (3.8) is the following. The fraction of secret bits shared by Alice and Bob per round is quantified by the amount of information that their raw key bits have in common $H(R_A : R_B)$ minus the information that Eve gained on Alice's key bit $H(R_A : E)$.

We now compute explicitly the key rate in (3.9) for the BB84 protocol, in terms of the observed quantities E_Z and E_X. For simplicity, in the computation we consider an asymmetric version of the BB84 protocol where the raw key is only extracted from Z basis measurements, while the X outcomes are used for PE (together with a fraction of Z outcomes).

The entropies in the key rate expression are computed on the c.c.q. state $\rho_{R_A R_B E}$ resulting after Alice and Bob measured their qubit in the Z basis to generate the raw key bits R_A and R_B, respectively. Alice and Bob's projective measurements are represented by the quantum maps $\mathcal{E}_{R_A}$ and $\mathcal{E}_{R_B}$ such that the state $\rho_{R_A R_B E}$ reads:

$$\rho_{R_A R_B E} = (\mathcal{E}_{R_A} \otimes \mathcal{E}_{R_B} \otimes \mathbb{1}_E)|\phi_{ABE}\rangle\langle\phi_{ABE}|$$

$$= \sum_{a,b=0}^{1} (P_{|a\rangle} \otimes P_{|b\rangle} \otimes \mathbb{1}_E)|\phi_{ABE}\rangle\langle\phi_{ABE}|(P_{|a\rangle} \otimes P_{|b\rangle} \otimes \mathbb{1}_E), \tag{3.10}$$

where $P_{|a\rangle} = |a\rangle\langle a|$ and similarly $P_{|b\rangle}$ are rank-one projectors on the Z basis, i.e. $|a\rangle, |b\rangle \in \{|0\rangle, |1\rangle\}$. We remark that we restricted without loss of generality to collective attacks where $|\phi_{ABE}\rangle$ represents the global state in a generic round of the protocol.

We start the key rate computation by assuming without loss of generality (w.l.o.g.) that, before distributing the state ρ_{AB} to the parties, Eve applies to it the maps $\mathcal{D}_1$ and $\mathcal{D}_2$, defined by:

$$\mathcal{D}_i(\rho_{AB}) = \frac{1}{2}\rho_{AB} + \frac{1}{2}D_i\rho_{AB}D_i^\dagger \quad i = 1, 2, \tag{3.11}$$

where the operators D_i read:

$$D_1 = X \otimes X \quad ; \quad D_2 = Z \otimes Z. \tag{3.12}$$

One can easily verify that the resulting state $\tilde{\rho}_{AB}$ received by Alice and Bob:

$$\tilde{\rho}_{AB} = (\mathcal{D}_1 \circ \mathcal{D}_2) \rho_{AB} = \frac{1}{4} [\rho_{AB} + (Z \otimes Z)\rho_{AB}(Z \otimes Z)$$
$$+ (X \otimes X)\rho_{AB}(X \otimes X) + (Y \otimes Y)\rho_{AB}(Y \otimes Y)] \tag{3.13}$$

is diagonal in the Bell basis $\{|\psi_{ij}\rangle\}^1_{i,j=0}$ of two qubits, with the same diagonal coefficients of the original state ρ_{AB}. We can thus express $\tilde{\rho}_{AB}$ in the Bell basis as:

$$\tilde{\rho}_{AB} = \sum_{i,j=0}^{1} \lambda_{ij} |\psi_{ij}\rangle\langle\psi_{ij}| \tag{3.14}$$

for some eigenvalues $0 \le \lambda_{ij} \le 1$ such that $\sum_{i,j} \lambda_{ij} = 1$, where the states of the Bell basis read:

$$|\psi_{ij}\rangle = \frac{|0, j\rangle + (-1)^i |1, 1 - j\rangle}{\sqrt{2}}, \quad i, j \in \{0, 1\}. \tag{3.15}$$

The assumption that Alice and Bob are given the Bell-diagonal state (3.14) is not restrictive due to two reasons. First, since the state $\tilde{\rho}_{AB}$ is prepared by Eve, she also holds its purification and one can show that her uncertainty on Alice's key is not increased when she distributes $\tilde{\rho}_{AB}$ in place of ρ_{AB}: $H(R_A|E)_\rho \ge H(R_A|E)_{\tilde{\rho}}$. The interested reader can find the proof of this fact in the Appendix of this Chapter (Sect. 3.5). The second reason is that from the point of view of the parties, the action of $\mathcal{D}_1 \circ \mathcal{D}_2$ corresponds to a simultaneous flip of both Alice's and Bob's bits, which occurs with probability $1/2$. This implies that the marginal distributions of Alice's and Bob's raw key bits are symmetrized. However, the observed QBERs are unaffected[4] as well as the correlation of the raw keys of Alice and Bob. Therefore, the only visible effect is the symmetrization of the marginals. This could be directly enforced by the parties by agreeing on flipping their outcomes with probability $1/2$, while communicating over the public channel. Thus Eve would be aware of the flipping.

For the above arguments, Eve distributes w.l.o.g. the state (3.14) to Alice and Bob.

Recall the definitions of the QBERs E_Z and E_X in terms of probabilities: The QBER E_Z (E_X) is the probability that the Z (X) measurement outcomes of Alice and Bob differ. Given that the parties share the state $\tilde{\rho}_{AB}$ in (3.14), it holds:

$$E_Z = \mathrm{Tr}[(P_{|0\rangle} \otimes P_{|1\rangle} + P_{|1\rangle} \otimes P_{|0\rangle})\tilde{\rho}_{AB}] = \lambda_{01} + \lambda_{11} \tag{3.16}$$
$$E_X = \mathrm{Tr}[(P_{|+\rangle} \otimes P_{|-\rangle} + P_{|-\rangle} \otimes P_{|+\rangle})\tilde{\rho}_{AB}] = \lambda_{10} + \lambda_{11}. \tag{3.17}$$

---

[4]This is due to the fact that either both Alice's and Bob's bits are flipped, or none is.

Moreover, Eve holds the purifying system of $\tilde{\rho}_{AB}$ such that the global pure state reads:

$$|\phi_{ABE}\rangle = \sum_{i,j=0}^{1} \sqrt{\lambda_{ij}} |\psi_{ij}\rangle_{AB} \otimes |e_{ij}\rangle_E, \qquad (3.18)$$

where $\{|e_{ij}\rangle\}_{i,j=0}^{1}$ is an orthonormal basis in $\mathcal{H}_E$.

In order to compute the conditional entropy $H(R_A|E)$, we express it as follows:

$$H(R_A|E) = H(E|R_A) + H(R_A) - H(E). \qquad (3.19)$$

The first term is computed on the state $\rho_{R_A E}$ derived from (3.10) by tracing out Bob's subsystem:

$$
\begin{aligned}
\rho_{R_A E} &= \sum_{a=0}^{1} |a\rangle\langle a|_{R_A} \otimes \mathrm{Tr}_{AB}\left[(|a\rangle\langle a| \otimes \mathbb{1}_{BE})|\phi_{ABE}\rangle\langle\phi_{ABE}|\right] \\
&= \sum_{a=0}^{1} |a\rangle\langle a|_{R_A} \otimes \sum_{i,j,k,l=0}^{1} \sqrt{\lambda_{ij}\lambda_{kl}}\,\mathrm{Tr}_{AB}\left[(|a\rangle\langle a| \otimes \mathbb{1}_B)|\psi_{ij}\rangle\langle\psi_{kl}|\right]|e_{ij}\rangle\langle e_{kl}|_E \\
&=: \sum_{a=0}^{1} \mathrm{Pr}(a)\,|a\rangle\langle a|_{R_A} \otimes \rho_E^a, \qquad (3.20)
\end{aligned}
$$

where the probability of Alice observing outcome a is $\mathrm{Pr}(a) = 1/2$ due to the symmetrized distribution of R_A, whereas Eve's state ρ_E^a, conditioned on Alice observing a, simplifies to:

$$\rho_E^a = \sum_{i,j,k=0}^{1} \sqrt{\lambda_{ij}\lambda_{kj}}\,(-1)^{(i+k)a}\,|e_{ij}\rangle\langle e_{kj}|. \qquad (3.21)$$

The non-zero eigenvalues of (3.21) are independent of a and given by: $\{\lambda_{00} + \lambda_{10}, \lambda_{01} + \lambda_{11}\}$. By recalling the expression (2.52) of the conditional entropy of a c.q. state, we can compute the first term in (3.19) as follows:

$$H(E|R_A) = \sum_{a=0}^{1} \mathrm{Pr}(a)H(\rho_E^a) = H(\{\lambda_{00} + \lambda_{10}, \lambda_{01} + \lambda_{11}\}) = h(E_Z), \qquad (3.22)$$

where we used the binary entropy $h(p)$ expression (2.46) and the fact that the coefficients λ_{ij} sum to one. Symmetrized marginals imply that $H(R_A) = 1$ and since the state on ABE is pure, the entropies of the subsystems E and AB are equal: $H(E) = H(AB) = H(\{\lambda_{ij}\})$. Substituting everything in (3.19) we obtain:

$$H(R_A|E) = 1 + h(E_Z) - H(\{\lambda_{ij}\}). \qquad (3.23)$$

Note that the eigenvalues $\{\lambda_{ij}\}$ are not completely fixed by the observed error rates E_Z and E_X through (3.16) and (3.17). Thus we must consider the worst-case scenario and minimize (3.23) over $\{\lambda_{ij}\}$, with the constraints given by (3.16) and (3.17). The minimization leads to the following result [6]:

$$H(R_A|E) = 1 + h(E_Z) - (h(E_X) + h(E_Z)) = 1 - h(E_X). \tag{3.24}$$

We remark that we could minimize $H(R_A|E)$ independently of $H(R_A|R_B)$ since the latter is fixed by the QBER E_Z. Indeed, the conditional Shannon entropy $H(R_A|R_B)$ is computed on the probability distribution $\Pr(a, b)$ of Alice and Bob's Z outcomes. Due to the symmetrization of the marginals, it is easy to express the entropy exclusively in terms of E_Z as follows:

$$H(R_A|R_B) = h(E_Z). \tag{3.25}$$

By employing (3.24) and (3.25) in (3.9), we obtain the asymptotic key rate of the BB84 protocol in terms of the observed error rates:

$$r_{\text{BB84}} = 1 - h(E_X) - h(E_Z). \tag{3.26}$$

3.3 Finite-Key Security

The asymptotic secret key rate given in (3.9) is only valid in the limit of infinitely many protocol rounds. Here, we generalize that result by presenting the secret key length achieved by a general QKD protocol with finite resources and prove its security. In doing so, we mainly follow the reference [12].

3.3.1 General QKD Protocol

Consider two parties, Alice and Bob, who have access to fresh randomness and are linked by an insecure quantum channel and an authenticated classical public channel (see Fig. 3.1). A potential eavesdropper, Eve, is assumed to have full control over the quantum channel and access to the messages sent via the public channel.

The parties run a QKD protocol, whose goal is to output a pair of identical keys (s_A, s_B) for Alice and Bob, respectively, completely unknown to Eve. The protocol could also abort and output the symbol: $s_A = s_B = \perp$. We describe the general QKD protocol in the entanglement-based view.

1. The protocol starts with the distribution of M quantum signals through the quantum channel. The joint state of the signals is represented by ρ_{AB}^M and Eve holds its purifying system (we allow for coherent attacks). Alice and Bob perform

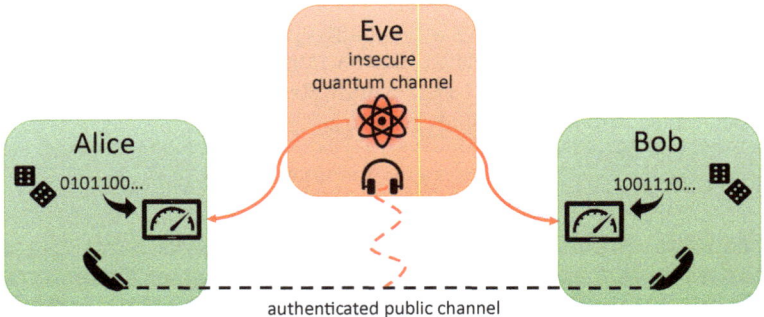

Fig. 3.1 Schematic representation of the setup of an entanglement-based QKD protocol, from a security perspective. In each round, Eve may distribute a quantum signal to Alice and Bob through the quantum channel. Alice and Bob locally measure the incoming signal with a randomly-chosen measurement setting and record the classical output. After the transmission of quantum signals is over, the parties communicate via the classical public channel to perform error correction and privacy amplification

local measurements on each signal received and collect the classical outcomes. Depending on the protocol, Alice and Bob can randomly choose among certain measurement settings. Typically, one setting is chosen with higher probability and is used for key generation, while the other(s) form the test rounds.

2. In PE, the parties reveal the settings and the outcomes of the test rounds, as well as the outcomes of a random sample of key-generation rounds. This information is used to estimate the noise in the quantum channel (and thus Eve's knowledge). If the noise is above a certain threshold, the protocol aborts.

3. At this point, both Alice and Bob hold a string of $n < M$ partially correlated key bits forming their raw key, denoted R_A^n and R_B^n, respectively. The parties perform an EC procedure in order for Bob to compute a guess $\hat{R}_A^n$ of Alice's raw key. In doing so, they reveal leak_{EC} bits of information over the public channel. In order to verify if EC was successful, Alice computes a hash h_A (bitstring) of length $\lceil \log(1/\varepsilon_{\text{EC}}) \rceil$ from her raw key R_A^n by applying a randomly-picked two-universal hash function (Definition 2.11). She publicly announces the hash function and h_A. Bob uses Alice's hash function to compute the hash h_B from his guess $\hat{R}_A^n$. If $h_A \neq h_B$, the protocol aborts. The total amount of information about Alice's raw key R_A^n revealed during EC is thus given by: $\text{leak}_{\text{EC}} + \lceil \log(1/\varepsilon_{\text{EC}}) \rceil \leq \text{leak}_{\text{EC}} + \log(2/\varepsilon_{\text{EC}})$.

4. In PA, Alice randomly picks another two-universal hash function and communicates it to Bob over the public channel. Both Alice and Bob apply the two-universal hash function to their error-corrected keys R_A^n and $\hat{R}_A^n$ and obtain shorter, secret keys s_A and s_B of length ℓ. The final key length ℓ is chosen such that:

$$\ell \leq H_{\min}^{\varepsilon}(R_A^n | E) - \text{leak}_{\text{EC}} - \log \frac{2}{\varepsilon_{\text{EC}}} - 2 \log \frac{1}{2\,\varepsilon_{\text{PA}}}, \tag{3.27}$$

for some $\varepsilon, \varepsilon_{\text{EC}}, \varepsilon_{\text{PA}} > 0$ which depend on the required level of security (see Subsect. 3.3.2).

> The length of the secret key (3.27) is determined by the smooth min-entropy of Alice's raw key, as discussed in Sect. 2.10, from which one subtracts the information leaked to Eve during EC.

The (non-asymptotic) secret key rate of the described protocol is given by:

$$r = \tau \frac{\ell}{M}, \qquad (3.28)$$

where τ is the repetition rate of the experimental setup, i.e. the inverse of the time needed to perform one round of the protocol (distribution of quantum signal and measurements). In this book we always consider $\tau = 1$.

Remark 3.1 *(Min-entropy estimation) We emphasize that the secret key length in (3.27) is valid for an arbitrary QKD protocol. However, the smooth min-entropy term appearing in its expression cannot be directly computed since Eve's action is unknown, i.e. the state $\rho^n_{R_A E}$ representing Alice's raw key and Eve's quantum side information is not known. Hence, the challenge of every QKD protocol is to estimate the min-entropy term in the tightest way possible, by relying on the observed data employed for PE.*

3.3.2   Security Definition and Proof

We now define what it means for a QKD protocol to be "secure" and subsequently prove the security of the general QKD protocol outlined above.

Definition 3.1 *(Correctness)* A QKD protocol is said to be ε_{cor}-correct if:

$$\Pr[s_A \neq s_B] \leq \varepsilon_{\text{cor}}. \qquad (3.29)$$

Definition 3.2 *(Secrecy)* A QKD protocol is said to be ε_{sec}-secret if, for Ω being the event that the protocol does not abort, the following inequality holds:

$$\Pr[\Omega] \, T\left(\rho_{S_A E_{\text{tot}} | \Omega}, \omega_{S_A} \otimes \rho_{E_{\text{tot}} | \Omega}\right) \leq \varepsilon_{\text{sec}}, \qquad (3.30)$$

where $\rho_{S_A E_{\text{tot}} | \Omega}$ is the state that describes the correlation between Alice's final secret key S_A and the total information available to Eve E_{tot} given that the protocol did not abort, while $\omega_{S_A} = \frac{1}{|S|} \sum_{s \in S} |s\rangle\langle s|$ is the maximally mixed state over all the possible realizations of Alice's key and $T(\cdot, \cdot)$ is the trace distance (Definition 2.13).

The correctness definition implies that the protocol always outputs identical keys for Alice and Bob, except for probability at most $\varepsilon_{\mathrm{cor}}$.

The secrecy statement is a bit more involved. A *real* QKD protocol is $\varepsilon_{\mathrm{sec}}$-secret if it is $\varepsilon_{\mathrm{sec}}$-indistinguishable from an *ideal* QKD protocol. By definition, the ideal QKD protocol acts exactly like the real protocol but it always outputs a perfectly secret key for Alice, i.e. a uniformly distributed key independent of Eve's knowledge, whenever the real protocol does not abort [7, 13]. This is formalized by stating that the output states of the real and ideal protocol are given by (we ignore Bob's system):

$$\rho^{\mathrm{real}}_{S_A E_{\mathrm{tot}}} = \Pr\left[\overline{\Omega}\right] \mid \bot\rangle\langle\bot \mid_{S_A} \otimes \rho_{E_{\mathrm{tot}}\mid\overline{\Omega}} + \Pr[\Omega]\,\rho_{S_A E_{\mathrm{tot}}\mid\Omega}$$
$$\rho^{\mathrm{ideal}}_{S_A E_{\mathrm{tot}}} = \Pr\left[\overline{\Omega}\right] \mid \bot\rangle\langle\bot \mid_{S_A} \otimes \rho_{E_{\mathrm{tot}}\mid\overline{\Omega}} + \Pr[\Omega]\,\omega_{S_A} \otimes \rho_{E_{\mathrm{tot}}\mid\Omega}, \qquad (3.31)$$

where $\overline{\Omega}$ is the event that the protocol aborts. The real and ideal protocol are $\varepsilon_{\mathrm{sec}}$-indistinguishable if their output states (3.31) are such,[5] i.e. when (Definition 2.14):

$$T(\rho^{\mathrm{real}}_{S_A E_{\mathrm{tot}}}, \rho^{\mathrm{ideal}}_{S_A E_{\mathrm{tot}}}) \leq \varepsilon_{\mathrm{sec}}. \qquad (3.32)$$

One can easily verify that (3.32) reduces to the condition (3.30) by using the definition of trace distance (Definition 2.13), since the abortion component of the two states in (3.31) cancels out when taking the difference.

The secrecy of Alice's key s_A alone does not guarantee that even Bob's key s_B is secret, unless we combine it with a statement on the correctness of the protocol. Therefore we define the security of a QKD protocol as follows.

Definition 3.3 (*Security*) A QKD protocol is said to be $\varepsilon_{\mathrm{tot}}$-secure if it is $\varepsilon_{\mathrm{cor}}$-correct and $\varepsilon_{\mathrm{sec}}$-secret, with $\varepsilon_{\mathrm{tot}} \geq \varepsilon_{\mathrm{cor}} + \varepsilon_{\mathrm{sec}}$.

Note that a trivial protocol that always aborts and outputs $s_A = s_B = \bot$ is secure according to the above definitions. Thus, another important feature of a QKD protocol is its *completeness*, i.e. the existence of an honest implementation of the protocol such that the probability of not aborting is $\Pr[\Omega] \geq 1 - \varepsilon_c$, for some small ε_c.

We also remark that the Definitions 3.1, 3.2 and 3.3 are *composable*. This means that when a QKD protocol—proven secure according to these definitions—is composed with another cryptographic task, the security of their combination can be inferred based on their individual security proofs and does not require a separate new proof. This is particularly relevant for QKD, which is often composed with one-time pads as discussed in Sect. 3.1.

Lemma 3.1 (Security of QKD) *The general QKD protocol of Sect. 3.3.1 is $\varepsilon_{\mathrm{tot}}$-secure, with $\varepsilon_{\mathrm{tot}} \geq \varepsilon_{\mathrm{EC}} + 2\varepsilon + \varepsilon_{\mathrm{PA}}$.*

---

[5]More precisely, the two protocols are $\varepsilon_{\mathrm{sec}}$-indistinguishable if the distinguishing advantage of an unbounded distinguisher, attempting to distinguish the real and ideal protocol, is upper bounded by $\varepsilon_{\mathrm{sec}}$. In [13] the authors show that such distinguishing advantage reduces to the trace distance between the output states of the protocols.

In order to prove this statement, one first shows that the general QKD protocol described earlier is ε_{EC}-correct. This is guaranteed by the fact that Alice and Bob verify the success of EC by computing and comparing hashes of length $\lceil \log(1/\varepsilon_{EC}) \rceil$. The second step is to show that the protocol is at least $(2\varepsilon + \varepsilon_{PA})$-secret by employing the Quantum Leftover Hash Lemma (c.f. Lemma 2.1), which is at the core of finite-key QKD security. We provide the full proof of Lemma 3.1 in the Appendix of this Chapter (Sect. 3.6).

3.3.3 Reduction to Asymptotic Key Rate

We emphasize that the non-asymptotic secret key rate in (3.28), computed with the key length in (3.27) of a generic QKD protocol, reduces to the asymptotic key rate given in (3.9) in the limit $M \to \infty$ of infinitely many rounds. This fact shows that the results presented in this Section properly generalize QKD key rates to the scenario of finite resources.

In order to prove the reduction of (3.28) to (3.9), we make use of an important tool called the *postselection technique* (PST) [14], valid for discrete-variable QKD protocols where the dimension $d = \dim(\mathcal{H}_A \otimes \mathcal{H}_B)$ of the quantum systems held by Alice and Bob can be characterized. The PST states that if a QKD protocol of M rounds is ε_{tot}-secure against collective attacks, then it is also $(M+1)^{d^2-1}\varepsilon_{tot}$-secure against coherent attacks if the secret key length (3.27) (the output of PA) is shortened by $2(d^2 - 1)\log(M + 1)$ bits.

Recall that in case of collective attacks, the state shared by the parties in the M rounds is the i.i.d. state $\rho_{AB}^{\otimes M}$, while for coherent attacks—as we consider in this finite-key analysis—the shared state is the more general ρ_{AB}^M.

Since in the asymptotic limit ($M \to \infty$, and $\varepsilon_{tot} \to 0$ exponentially fast) the corrections to the secret key rate introduced by the PST are negligible, proving the security of a generic QKD protocol against coherent attacks reduces to proving the security of the same protocol against collective attacks [6, 14, 15]. In other words, we can assume without loss of generality that the state distributed to the parties by Eve is an i.i.d. state $\rho_{AB}^{\otimes M}$. As a consequence, the smooth min-entropy term in (3.27) is now computed on the state $\rho_{R_A E}^{\otimes n}$: $H_{\min}^{\varepsilon}(R_A^n | E)_{\rho_{R_A E}^{\otimes n}}$.

Moreover, by recalling the operational meaning of the smooth max-entropy (c.f. Sect. 2.6), the minimum amount of leakage in (3.27) is quantified by $\text{leak}_{EC} \approx H_{\max}^{\varepsilon'}(R_A^n | R_B^n)$, where we neglected terms that tend to zero in the asymptotic limit. In case of collective attacks, the smooth-max entropy is evaluated on the i.i.d. state $\rho_{R_A R_B}^{\otimes n}$ and reads: $H_{\max}^{\varepsilon'}(R_A^n | R_B^n)_{\rho_{R_A R_B}^{\otimes n}}$.

Finally, by applying the AEP (2.63) and (2.64) on the smooth entropy terms appearing in (3.27), we reduce them to the correspondent von Neumann entropies: $H(R_A|E)$ and $H(R_A|R_B)$. In this way (3.9) is recovered.

Note that the PST has been fundamental for the application of the AEP, since the latter only holds for i.i.d. quantum states.

3.4 State-of-the-Art Experiments

In this Section we provide a brief and incomplete overview of the most recent experimental achievements in QKD. More complete and elaborated reviews can be found in [6, 7, 16].

In October 1989 Bennett, Brassard and other scientists implemented for the first time a QKD protocol, specifically the BB84 protocol [17, 18]. The experiment was carried out in a laboratory and was characterized by the transmission of polarized light over (just) 32.5 cm.

Since then much progress has been made, also thanks to the interest and investments of governments and companies [19, 20]. Current QKD implementations can reach secret key rates of the order of Mbits^{-1} over about 50 km of telecom fibre [21–23]. By exploiting wavelength division multiplexing (WDM), scientists have multiplexed 194 QKD channels over the same fibre reaching an aggregate key rate of 172.6 Mbits^{-1} [24].

In view of implementing QKD in existing optical networks, it has been demonstrated that QKD can be multiplexed alongside classical data channels amounting to a total of 18.3 Tbit s^{-1} of classical datarate [25], which is the typical datarate achieved nowadays in telecom fibres. To the same aim, QKD has successfully undergone field tests on commercial telecom fibres [26–28].

Thanks to novel architectures and security proofs, namely measurement-device-independent QKD (Chap. 5) and twin-field QKD (Chap. 6), it has been possible to extend the achievable distance of QKD on telecom fibres to over 400 km [29–32], with the record currently being set to 509 km by a twin-field QKD protocol [33].

By using free-space optical links, such as satellite-to-ground links, it is possible to extend the maximum distance of QKD even further. In 2017, Chinese and Japanese research groups independently realized the first QKD protocols in free-space using low-Earth-orbit satellites [34, 35]. In particular, the Chinese group led by Prof. Pan implemented QKD over 1000 km between the satellite and a ground station [34], including a quantum-secured video call between Beijing and Vienna [36]. More recently, Prof. Pan's group performed entanglement-based QKD between two ground stations separated by 1120 km using a satellite as the source of the entangled states [37].

Appendix

In this Appendix we prove some statements made in the main text, whose articulated proof would have altered the cohesion and flow of the text.

3.5 Eve's Uncertainty Is Non-increasing Under Symmetrization

Part of the proof in this Section is inspired by [38].

In computing the secret key rate of the BB84 protocol [5] in Sect. 3.2, we argue that w.l.o.g. the state ρ_{AB} distributed to Alice and Bob by Eve is replaced by (3.13):

$$\tilde{\rho}_{AB} = \frac{1}{4}[\rho_{AB} + (Z \otimes Z)\rho_{AB}(Z \otimes Z) + (X \otimes X)\rho_{AB}(X \otimes X)$$
$$+(Y \otimes Y)\rho_{AB}(Y \otimes Y)]. \tag{3.33}$$

This scenario can be viewed as Eve preparing one of the four states

$$\rho_{AB}, \quad (Z \otimes Z)\rho_{AB}(Z \otimes Z), \quad (X \otimes X)\rho_{AB}(X \otimes X), \quad (Y \otimes Y)\rho_{AB}(Y \otimes Y) \tag{3.34}$$

depending on the outcome $t = 1, 2, 3, 4$ of a random variable stored in the register T, which Eve is aware of. Since Eve holds the purifying system E of every state in (3.34), the state prepared by Eve is:

$$\tilde{\rho}_{ABET} = \frac{1}{4}\sum_t |\phi^t_{ABE}\rangle\langle\phi^t_{ABE}| \otimes |t\rangle\langle t|_T, \tag{3.35}$$

where $\{|\phi^t_{ABE}\rangle\}^4_{t=1}$ are pure states. Finally, we assume that Eve holds the purifying system T' of the state in (3.35). Thus the global state is pure and reads:

$$|\phi_{ABETT'}\rangle = \frac{1}{2}\sum_t |\phi^t_{ABE}\rangle \otimes |t\rangle_T \otimes |t\rangle_{T'}. \tag{3.36}$$

Note that (3.36) is a purification of (3.35), where both registers T and T' are held by Eve. The above argument holds only if it's not disadvantageous for Eve. In other words, Eve's uncertainty on Alice's key, quantified by the conditional entropy $H(R_A|E)$, must be non-increasing. Therefore, we must verify that:

$$H(R_A|E)_\rho \geq H(R_A|E_{\text{tot}})_{\tilde{\rho}}, \tag{3.37}$$

where Eve's quantum system $E_{\text{tot}} = ETT'$ contains: the quantum side information E, the outcome of the random variable T, and the purifying system T'.

Proof In order to prove (3.37), we start by using the strong subadditivity property (c.f. Sect. 2.6):

$$H(R_A|E_{\text{tot}})_{\tilde{\rho}} \leq H(R_A|ET)_{\tilde{\rho}} \tag{3.38}$$

where the r.h.s. entropy is computed on the following state:

$$
\begin{aligned}
\tilde{\rho}_{R_A ET} &= (\mathcal{E}_{R_A} \otimes \mathrm{id}_{ET}) \, \mathrm{Tr}_B \left[\tilde{\rho}_{ABET} \right] \\
&= \frac{1}{4} (\mathcal{E}_{R_A} \otimes \mathrm{id}_{ET}) \, \mathrm{Tr}_B \left[\sum_t |\phi^t_{ABE}\rangle\langle\phi^t_{ABE}| \otimes |t\rangle\langle t|_T \right] \\
&=: \frac{1}{4} \sum_t \rho^t_{R_A E} \otimes |t\rangle\langle t|_T,
\end{aligned}
\tag{3.39}
$$

where the quantum map

$$
\mathcal{E}_{R_A}(\sigma) = \sum_{a=0}^{1} |a\rangle\langle a| \, \langle a|\sigma|a\rangle
$$

represents the measurement performed by Alice for key generation, i.e. a projection onto the Z basis. Being the state in Eq. (3.39) a c.q. state, its entropy simplifies to:

$$
H(R_A|ET)_{\tilde{\rho}} = \frac{1}{4} \sum_t H(R_A|E)_{\rho^t}.
\tag{3.40}
$$

The last part of the proof shows that $H(R_A|E)_{\rho^t}$ is actually independent of t and equal to conditional entropy of the original state $H(R_A|E)_\rho$. This is clear if the state $\rho^t_{R_A E}$ is made explicit. From Eq. (3.39) we have that:

$$
\rho^t_{R_A E} = (\mathcal{E}_{R_A} \otimes \mathrm{id}_{ET}) \, \mathrm{Tr}_B \left[|\phi^t_{ABE}\rangle\langle\phi^t_{ABE}| \right],
\tag{3.41}
$$

where $|\phi^t_{ABE}\rangle$ is the purification of one of the four states in (3.34) prepared by Eve according to the random variable T. For definiteness, let's fix that state to be $(X \otimes X) \rho_{AB} (X \otimes X)$, although an analogous reasoning holds for any other state in Eq. (3.34). By writing ρ_{AB} in its spectral decomposition:

$$
\rho_{AB} = \sum_\lambda \lambda |\lambda\rangle\langle\lambda|,
\tag{3.42}
$$

we can immediately explicit $|\phi^t_{ABE}\rangle$ as follows:

$$
|\phi^t_{ABE}\rangle = \sum_\lambda \sqrt{\lambda} |\lambda'\rangle_{AB} \otimes |e_\lambda\rangle_E,
\tag{3.43}
$$

where the eigenstates of the operator $(X \otimes X) \rho_{AB} (X \otimes X)$ read: $|\lambda'\rangle = (X \otimes X)|\lambda\rangle$. By substituting (3.43) into (3.41) and by making explicit the map $\mathcal{E}_{R_A}$ we obtain the following chain of equalities:

$$
\begin{aligned}
\rho_{R_A E}^t &= \sum_{a=0}^{1} |a\rangle\langle a| \otimes \sum_{\lambda,\sigma} \sqrt{\lambda\sigma}\, \mathrm{Tr}_B\left[\langle a||\lambda'\rangle\langle\sigma'||a\rangle\right] |e_\lambda\rangle\langle e_\sigma| \\
&= \sum_{a=0}^{1} |a\rangle\langle a| \otimes \sum_{\lambda,\sigma} \sqrt{\lambda\sigma}\, \mathrm{Tr}_B\left[\langle a|(X\otimes X)|\lambda\rangle\langle\sigma|(X\otimes X)|a\rangle\right] |e_\lambda\rangle\langle e_\sigma| \\
&= \sum_{a=0}^{1} |a\rangle\langle a| \otimes \sum_{\lambda,\sigma} \sqrt{\lambda\sigma}\, \mathrm{Tr}_B\left[\langle\bar{a}||\lambda\rangle\langle\sigma||\bar{a}\rangle\right] |e_\lambda\rangle\langle e_\sigma| \\
&= \sum_{a=0}^{1} |\bar{a}\rangle\langle\bar{a}| \otimes \sum_{\lambda,\sigma} \sqrt{\lambda\sigma}\, \mathrm{Tr}_B\left[\langle a||\lambda\rangle\langle\sigma||a\rangle\right] |e_\lambda\rangle\langle e_\sigma| \\
&=: \sum_{a=0}^{1} |\bar{a}\rangle\langle\bar{a}| \otimes \rho_E^a,
\end{aligned}
\tag{3.44}
$$

where in the third equality we used the cyclic property of the trace and the fact that Alice measures in the Z basis $\{|0\rangle, |1\rangle\}$, hence the Pauli operator X flips its eigenstates: $X|a\rangle = |\bar{a}\rangle$. In the fourth equality we relabelled the classical outcomes: $a \leftrightarrow \bar{a}$. Finally, by comparing (3.44) with the state $\rho_{R_A E}$ obtained in an analogous way from the original state ρ_{AB}:

$$
\begin{aligned}
\rho_{R_A E} &= (\mathcal{E}_{R_A} \otimes \mathbb{1}_E)\, \mathrm{Tr}_B\left[|\phi_{ABE}\rangle\langle\phi_{ABE}|\right] \\
&= \sum_{a=0}^{1} |a\rangle\langle a| \otimes \sum_{\lambda,\sigma} \sqrt{\lambda\sigma}\, \mathrm{Tr}_B\left[\langle a||\lambda\rangle\langle\sigma||a\rangle\right] |e_\lambda\rangle\langle e_\sigma| \\
&= \sum_{a=0}^{1} |a\rangle\langle a| \otimes \rho_E^a,
\end{aligned}
\tag{3.45}
$$

we observe that $\rho_{R_A E}^t$ and $\rho_{R_A E}$ are the same state up to a permutation of the classical outcomes, thus their conditional entropies coincide:

$$
H(R_A|E)_{\rho^t} = H(R_A|E)_\rho \quad \forall t.
\tag{3.46}
$$

In conclusion, by combining Eqs. (3.46), (3.40) and (3.38), we prove the claim in Eq. (3.37). This concludes the proof. $\qquad\square$

3.6 Finite-Key Security of QKD

Here we prove Lemma 3.1, following the lines of [9, 12].

Proof We start by showing that the protocol is $\varepsilon_{\mathrm{EC}}$-correct.

Recall that at the end of EC, Alice and Bob apply a two-universal hash function on their raw keys R_A^n and $\hat{R}_A^n$, obtaining hashes h_A and h_B of length $\lceil \log(1/\varepsilon_{\text{EC}}) \rceil$. The defining feature of two-universal hash functions is that the probability that two outputs of length $\lceil \log(1/\varepsilon_{\text{EC}}) \rceil$ coincide, given that the inputs are different, is small, namely: $2^{-\lceil \log(1/\varepsilon_{\text{EC}}) \rceil}$ (see Definition 2.11). In formulas, we have that:

$$\Pr[h_A = h_B, R_A^n \neq \hat{R}_A^n] \leq \Pr[h_A = h_B | R_A^n \neq \hat{R}_A^n] \leq 2^{-\lceil \log(1/\varepsilon_{\text{EC}}) \rceil} \leq \varepsilon_{\text{EC}}. \quad (3.47)$$

Then we observe that the keys s_A and s_B always coincide when the protocol aborts, thus $\Pr[s_A \neq s_B, h_A \neq h_B] = 0$. By employing (3.47) in the following expression, we prove that the protocol is ε_{EC}-correct:

$$\Pr[s_A \neq s_B] = \Pr[s_A \neq s_B, h_A = h_B] \leq \Pr[R_A^n \neq \hat{R}_A^n, h_A = h_B] \leq \varepsilon_{\text{EC}}. \quad (3.48)$$

In order to prove the secrecy, we make use of the Quantum Leftover Hash Lemma [9, 39], which provides the following upper bound:

$$\frac{1}{2} \left\| \rho_{S_A E_{\text{tot}} | \Omega} - \omega_{S_A} \otimes \rho_{E_{\text{tot}} | \Omega} \right\| \leq 2\varepsilon + \frac{1}{2} \sqrt{2^{\ell - H_{\min}^\varepsilon(R_A^n | CE)}}, \quad (3.49)$$

where ℓ is the length of Alice's key after PA and where we emphasize E_{tot} being the total information available to Eve. This comprises her purifying system E, the classical communication C occurred during EC and the knowledge F of the hash function used in PA: $E_{\text{tot}} = FCE$.

We now employ the following chain-rule for the min-entropy [12]:

$$H_{\min}^\varepsilon(R_A^n | CE) \geq H_{\min}^\varepsilon(R_A^n | E) - \log |C|$$
$$= H_{\min}^\varepsilon(R_A^n | E) - \text{leak}_{\text{EC}} - \log \frac{2}{\varepsilon_{\text{EC}}}, \quad (3.50)$$

where $\log |C|$ quantifies all the information revealed during EC and is given by $\text{leak}_{\text{EC}} + \log(2/\varepsilon_{\text{EC}})$ (see the protocol's description).

By inserting Eq. (3.50) into (3.49) we obtain the following chain of inequalities:

$$\frac{1}{2} \left\| \rho_{S_A E_{\text{tot}} | \Omega} - \omega_{S_A} \otimes \rho_{E_{\text{tot}} | \Omega} \right\| \leq 2\varepsilon + \frac{1}{2} \sqrt{2^{\ell - (H_{\min}^\varepsilon(R_A^n | E) - \text{leak}_{\text{EC}} - \log(2/\varepsilon_{\text{EC}}))}}$$
$$\leq 2\varepsilon + \frac{1}{2} \sqrt{2^{\log(2\varepsilon_{\text{PA}})^2}}$$
$$= 2\varepsilon + \varepsilon_{\text{PA}}, \quad (3.51)$$

where we used the key length expression (3.27) in the second inequality. We have thus proven that the protocol is ε_{sec}-secret, with $\varepsilon_{\text{sec}} \geq 2\varepsilon + \varepsilon_{\text{PA}}$. By combining this with the correctness proof, we have shown that the protocol is ε_{tot}-secure, with $\varepsilon_{\text{tot}} \geq 2\varepsilon + \varepsilon_{\text{PA}} + \varepsilon_{\text{EC}}$. This concludes the proof. $\qquad \square$

References

1. Miller, F. (1882). Telegraphic code to insure privacy and secrecy in the transmission of telegrams.
2. Vernam, G. S. (1926). Cipher printing telegraph systems for secret wire and radio telegraphic communications. *Transactions of the American Institute of Electrical Engineers, XLV*, 295–301.
3. Nielsen, M. A., & Chuang, I. L. (2010). *Quantum computation and quantum information* (10th Anniversary ed.). Cambridge University Press.
4. Wootters, W. K., & Zurek, W. H. (1982). A single quantum cannot be cloned. *Nature, 299*(5886), 802–803.
5. Bennett, C. H. and Brassard, G. (1984). Quantum cryptography: Public key distribution and coin tossing. In *Proceedings of IEEE International Conference on Computers, Systems and Signal Processing*, pp. 175 – 179.
6. Scarani, V., Bechmann-Pasquinucci, H., Cerf, N. J., Dušek, M., Lütkenhaus, N., & Peev, M. (2009). The security of practical quantum key distribution. *Reviews of Modern Physics, 81*, 1301–1350.
7. Pirandola, S., Andersen, U. L., Banchi, L., Berta, M., Bunandar, D., Colbeck, R., Englund, D., Gehring, T., Lupo, C., Ottaviani, C., Pereira, J., Razavi, M., Shaari, J. S., Tomamichel, M., Usenko, V. C., Vallone, G., Villoresi, P., & Wallden, P. (2019). Advances in quantum cryptography. arXiv:quant-ph/1906.01645.
8. Devetak, I., & Winter, A. (2005). Distillation of secret key and entanglement from quantum states. *Proceedings of the Royal Society, 461*.
9. Renner, R. (2008). Security of quantum key distribution. *International Journal of Quantum Information, 06*(01), 1–127.
10. Scarani, V., & Renner, R. (2008a). Quantum cryptography with finite resources: Unconditional security bound for discrete-variable protocols with one-way postprocessing. *Physical Review Letters, 100*, 200501.
11. Scarani, V., & Renner, R. (2008b). Security bounds for quantum cryptography with finite resources. In Kawano, Y., & Mosca, M., (eds.), *Theory of Quantum Computation, Communication, and Cryptography* (pp. 83–95). Springer, Berlin, Heidelberg.
12. Tomamichel, M., Lim, C. C. W., Gisin, N., & Renner, R. (2012). Tight finite-key analysis for quantum cryptography. *Nature Communications, 3*(1), 634.
13. Portmann, C., & Renner, R. (2014). Cryptographic security of quantum key distribution. arXiv:quant-ph/1409.3525.
14. Christandl, M., König, R., & Renner, R. (2009). Postselection technique for quantum channels with applications to quantum cryptography. *Physical Review Letters, 102*, 020504.
15. Renner, R. (2007). Symmetry of large physical systems implies independence of subsystems. *Nature Physics, 3*(9), 645–649.
16. Diamanti, E., Lo, H.-K., Qi, B., & Yuan, Z. (2016). Practical challenges in quantum key distribution. *npj Quantum Information, 2*(1), 16025.
17. Bennett, C. H., & Brassard, G. (1989). Experimental quantum cryptography: The dawn of a new era for quantum cryptography: The experimental prototype is working]. *SIGACT News, 20*(4), 78–80.
18. Bennett, C. H., Bessette, F., Brassard, G., Salvail, L., & Smolin, J. (1992). Experimental quantum cryptography. *Journal of Cryptology, 5*(1), 3–28.
19. Commission, E. The quantum flagship. https://qt.eu.
20. Technology, I. Q. Quantum key distribution (qkd) markets: 2019-2028. https://www.insidequantumtechnology.com/product/quantum-key-distribution-qkd-markets-2019-2028.
21. Dixon, A. R., Yuan, Z. L., Dynes, J. F., Sharpe, A. W., & Shields, A. J. (2008). Gigahertz decoy quantum key distribution with 1 mbit/s secure key rate. *Optics Express, 16*(23), 18790–18797.
22. Patel, K. A., Dynes, J. F., Lucamarini, M., Choi, I., Sharpe, A. W., Yuan, Z. L., et al. (2014). Quantum key distribution for 10 gb/s dense wavelength division multiplexing networks. *Applied Physics Letters, 104*(5), 051123.

23. Huang, D., Lin, D., Wang, C., Liu, W., Fang, S., Peng, J., et al. (2015). Continuous-variable quantum key distribution with 1 mbps secure key rate. *Optics Express*, *23*(13), 17511–17519.
24. Eriksson, T. A., Luís, R. S., Puttnam, B. J., Rademacher, G., Fujiwara, M., Awaji, Y., et al. (2020). Wavelength division multiplexing of 194 continuous variable quantum key distribution channels. *Journal of Lightwave Technology*, *38*(8), 2214–2218.
25. Eriksson, T. A., Hirano, T., Puttnam, B. J., Rademacher, G., Luís, R. S., Fujiwara, M., Namiki, R., Awaji, Y., Takeoka, M., Wada, N., & Sasaki, M. (2019). Wavelength division multiplexing of continuous variable quantum key distribution and 18.3 tbit/s data channels. *Communications Physics*, *2*(1), 9.
26. Zhang, Y., Li, Z., Chen, Z., Weedbrook, C., Zhao, Y., Wang, X., et al. (2019). Continuous-variable QKD over 50 km commercial fiber. *Quantum Science and Technology*, *4*(3), 035006.
27. Sasaki, M., Fujiwara, M., Ishizuka, H., Klaus, W., Wakui, K., Takeoka, M., et al. (2011). Field test of quantum key distribution in the Tokyo qkd network. *Optics Express*, *19*(11), 10387–10409.
28. Dynes, J. F., Wonfor, A., Tam, W. W.-S., Sharpe, A. W., Takahashi, R., Lucamarini, M., et al. (2019). Cambridge quantum network. npj Quantum. *Information*, *5*(1), 101.
29. Yin, H.-L., Chen, T.-Y., Yu, Z.-W., Liu, H., You, L.-X., Zhou, Y.-H., et al. (2016). Measurement-device-independent quantum key distribution over a 404 km optical fiber. *Physical Review Letters*, *117*, 190501.
30. Boaron, A., Boso, G., Rusca, D., Vulliez, C., Autebert, C., Caloz, M., et al. (2018). Secure quantum key distribution over 421 km of optical fiber. *Physical Review Letters*, *121*, 190502.
31. Wang, S., He, D.-Y., Yin, Z.-Q., Lu, F.-Y., Cui, C.-H., Chen, W., et al. (2019). Beating the fundamental rate-distance limit in a proof-of-principle quantum key distribution system. *Physical Review X*, *9*, 021046.
32. Liu, Y., Yu, Z.-W., Zhang, W., Guan, J.-Y., Chen, J.-P., Zhang, C., et al. (2019). Experimental twin-field quantum key distribution through sending or not sending. *Physical Review Letters*, *123*, 100505.
33. Chen, J.-P., Zhang, C., Liu, Y., Jiang, C., Zhang, W., Hu, X.-L., et al. (2020). Sending-or-not-sending with independent lasers: Secure twin-field quantum key distribution over 509 km. *Physical Review Letters*, *124*, 070501.
34. Liao, S.-K., Cai, W.-Q., Liu, W.-Y., Zhang, L., Li, Y., Ren, J.-G., et al. (2017). Satellite-to-ground quantum key distribution. *Nature*, *549*(7670), 43–47.
35. Takenaka, H., Carrasco-Casado, A., Fujiwara, M., Kitamura, M., Sasaki, M., & Toyoshima, M. (2017). Satellite-to-ground quantum-limited communication using a 50-kg-class microsatellite. *Nature Photonics*, *11*(8), 502–508.
36. Liao, S.-K., Cai, W.-Q., Handsteiner, J., Liu, B., Yin, J., Zhang, L., et al. (2018). Satellite-relayed intercontinental quantum network. *Physical Review Letters*, *120*, 030501.
37. Yin, J., Li, Y.-H., Liao, S.-K., Yang, M., Cao, Y., Zhang, L., et al. (2020). Entanglement-based secure quantum cryptography over 1,120 kilometres. *Nature*, *582*(7813), 501–505.
38. Watanabe, S., Matsumoto, R., Uyematsu, T., & Kawano, Y. (2007). Key rate of quantum key distribution with hashed two-way classical communication. *Physical Review A*, *76*, 032312.
39. Tomamichel, M., Schaffner, C., Smith, A., & Renner, R. (2011). Leftover hashing against quantum side information. *IEEE Transactions on Information Theory*, *57*(8), 5524–5535.

Chapter 4
Quantum Conference Key Agreement

Abstract Quantum conference key agreement (CKA) extends the notion of quantum key distribution (QKD) to the multipartite scenario. We introduce CKA in Sect. 4.1 and present the multipartite generalization of the BB84 protocol, including insight on its security proof and asymptotic key rate. In Sect. 4.2 we describe the functioning of a general CKA protocol and define its security, which is proven in Sect. 4.4 of the Appendix. We conclude the Chapter by discussing the first experimental implementations of CKA (Sect. 4.3).

The rapid development of quantum technologies allows us to foresee quantum networks [1–4] as one of its near-future applications. Quantum networks could be composed of matter-based quantum nodes where quantum information can be processed and stored, linked together by quantum channels where light distributes entangled states. Successful experiments on matter-light entanglement [5, 6] bring us closer to realizing such networks. The ultimate vision for quantum networks is building the quantum internet [7, 8].

A more accessible application of quantum networks is the generalization of the task of quantum key distribution (QKD) to a multiparty scenario, in what is called multipartite QKD or quantum conference key agreement (CKA). Here, N parties in a quantum network wish to establish a common secret key —a conference key— and use it to securely broadcast messages within the network. The first complete review on this topic is given in [9].

A CKA could be carried out by simply performing bipartite QKD schemes between pairs of parties, and then employing the established keys to securely distribute the conference key to all involved parties. However, such a solution would not exploit the possibility offered by quantum networks of distributing multipartite entangled states across several network nodes.

Conversely, it is possible to devise CKA protocols which make use of the correlations arising in multipartite entangled states in order to establish a conference key among several users [10–15]. This type of truly multipartite schemes can outperform the solution based on the iteration of bipartite schemes in certain network configurations (e.g. networks with bottlenecks) [11] and noise regimes [16]. It is worth

© The Author(s), under exclusive license to Springer Nature Switzerland AG 2021
F. Grasselli, *Quantum Cryptography*, Quantum Science and Technology,
https://doi.org/10.1007/978-3-030-64360-7_4

mentioning that the conference key rates achievable in a given network configuration are upper bounded by recently-derived fundamental limits, which depend on the network topology [17–20].

4.1   Extending QKD to Multiple Parties

In this Section we introduce the first discrete-variable CKA protocols by providing an intuitive explanation of their development.

Consider a scenario where Alice and $N - 1$ Bobs, denoted B_1, B_2 up to B_{N-1}, want to establish a secret conference key with a generalization of the BB84 protocol [21] presented in Sect. 3.2. In this multipartite scenario the conference key is extracted from Alice's raw key, hence during error correction every Bob attempts to correct his raw key to match Alice's. As a consequence, even in CKA protocols the main quantity to be estimated is the smooth min-entropy $H_{\min}^{\varepsilon}(R_A^n|E)$ of Alice's raw key conditioned on Eve's information (see Remark 3.1).

A naive approach to generalize the BB84 protocol would be to reproduce its prepare-and-measure description, where Alice now sends a state $|\phi_k\rangle$ ($k = 1, \ldots, 4$) out of the four states $\{|0\rangle, |1\rangle, |+\rangle, |-\rangle\}$ to every Bob ($|\pm\rangle = (|0\rangle \pm |1\rangle)/\sqrt{2}$). This means that in each round of the protocol the product state $|\phi_k\rangle^{\otimes(N-1)}$ is sent through the quantum channel. Since Eve is in control of the whole quantum channel, she can attempt to distinguish the four product states $|\phi_k\rangle^{\otimes(N-1)}$, whose overlap (inner product) is either 0 or $(1/\sqrt{2})^{N-1}$. As N increases, the four states become more distinguishable since their trace distance increases (c.f. Sect. 2.11), thus allowing Eve to retrieve more information about the key without being noticed. This leads to a dramatic decrease of the secret key rate eventually making the protocol useless, even assuming a flawless implementation.

The described CKA does not rely on entangled states (like the original BB84 protocol) and has actually been investigated for $N = 3$ in [22]. However, the idea is clearly not scalable to larger numbers of parties.

In order to devise a generalization of the BB84 protocol which would work with an arbitrary number of parties, we resort to its entanglement-based description. In the ideal implementation of the BB84 protocol, Alice and Bob measure their qubit in either the Z or X basis and obtain perfectly correlated and random outcomes since their qubits have been prepared in the Bell state $|\Phi^+\rangle$. Typically, the outcomes of the Z basis are used for key generation and those of the X basis are used to estimate the noise E_X in the channel and thus Eve's knowledge.

In generalizing this idea to $N \geq 3$ parties we encounter a fundamental problem. The only N-qubit state which leads to perfectly correlated and random outcomes in one measurement basis —necessary condition for generating a shared key— is the N-party GHZ state:

$$|\mathrm{GHZ}_N\rangle = \frac{1}{\sqrt{2}} \left[|0\rangle^{\otimes N} + |1\rangle^{\otimes N} \right], \tag{4.1}$$

when measured in the Z basis. However, the authors in [11] prove that even bipartite perfect correlations are forbidden in any other basis, contrary to what happens with the Bell state $|\Phi^+\rangle$ for $N = 2$.

Therefore, in an ideal N-party BB84 protocol Alice and the Bobs share an N-party GHZ state and measure in the Z basis for key generation.[1] However, they cannot estimate the channel's noise by a pairwise comparison of the X outcomes (or any other basis), since they would be uncorrelated even in the ideal scenario. How can we still estimate Eve's knowledge in the multipartite scenario?

Recall that the goal is to find a lower bound on the min-entropy $H_{\min}^{\varepsilon}(R_A^n|E)$, or on the von Neumann entropy $H(R_A|E)$ in the asymptotic scenario, of Alice's raw key given Eve's side information (c.f. Chap. 3).

One possible solution is provided in [11] for the asymptotic scenario. It basically consists in requiring the parties to measure their qubit in one of three bases, namely the X, Y or Z basis. In doing so, the parties can sufficiently characterize the state $\rho_{R_A E}$ describing Alice's raw key and Eve's quantum side information, to the extent that $H(R_A|E)$ is completely fixed by the measurement statistics. This solution can be interpreted as the N-party generalization of the *six-state QKD protocol* [23], where Alice and Bob are required to measure in the same three bases.

Alternatively, the security of a multipartite QKD scheme based on the GHZ state can also be ensured with just two measurement bases, making the protocol a multipartite version of the BB84 protocol. In this case, the parties measure in the Z basis for key generation and in the X basis to estimate Eve's knowledge of Alice's raw key [13]. We discuss in detail the multipartite BB84 protocol in the next Subsection.

In [13], the finite-key security of both the multipartite BB84 protocol and the multipartite six-state protocol introduced in [11] is proven,[2] and their performance is compared. As expected, the multipartite six-state protocol outperforms the multipartite BB84 protocol in the asymptotic limit of infinitely many rounds due to a more complete characterization of Eve's information. However, for lower number of rounds, the latter protocol provides higher secret key rates thanks to its tighter security analysis.

Remark 4.1 *(Entanglement is necessary) We emphasize that both the N-party six state protocol and the N-party BB84 protocol require the generation of entangled states even in their prepare-and-measure version. This contrasts with the bipartite BB84 protocol where Alice sends simple qubits to Bob.*

Indeed, in the entanglement-based view of the two multipartite QKD protocols, the parties are given the N-partite GHZ state (4.1). Now note that the conditional state of the Bobs, given that Alice measured X on the GHZ state and obtained outcome a (a = ±1), reads:

$$|\psi_a\rangle_{B_1...B_{N-1}} = \frac{1}{\sqrt{2}}\left(|0\rangle^{\otimes(N-1)} + a|1\rangle^{\otimes(N-1)}\right),\qquad(4.2)$$

---

[1] We remark that CKA is also possible with other resource states, such as the W state (Chap. 6).

[2] In Sect. 4.2 we rigorously define the security of CKA protocols by providing analogous definitions to those presented in Sect. 3.3.

which is an entangled state. Therefore, in an equivalent prepare-and-measure version of the protocol, Alice would need to prepare the entangled state (4.2) and send it to the Bobs in the rounds where she chooses the X basis.

However, the X-basis rounds are test rounds and are much less frequent than the key-generation rounds. In the key-generation rounds the conditional state of the Bobs, upon Alice measuring Z, is given by one of the two product states $|0\rangle^{\otimes(N-1)}$ and $|1\rangle^{\otimes(N-1)}$. Hence for key generation Alice can just prepare the same qubit state $N - 1$ times and send each of them to the corresponding Bob.

4.1.1 Multipartite BB84 Protocol

The multipartite BB84 protocol [13] allows N parties to establish a secret conference key by performing measurements in only two bases, the Z and the X basis.

The main idea is to view all the Bobs as one single Bob and define E_X as the error rate between the X outcomes of Alice (X_A) and the product of the X outcomes of all Bobs $(X_{\Pi B} := \prod_{i=1}^{N-1} X_{B_i})$, that is: $E_X = \Pr[X_A \neq X_{\Pi B}]$. Then, in an ideal implementation where the parties share the GHZ state (4.1), X_A and $X_{\Pi B}$ are perfectly correlated[3] like in the BB84 protocol and the channel noise is zero: $E_X = 0$. Therefore, when the parties observe an error rate $E_X \neq 0$, they conclude that Eve might have tampered with the quantum channel and can quantify the knowledge she acquired.

The N-party BB84 protocol comprises the following steps.

1. N qubits prepared in the GHZ state (4.1) are distributed to Alice and the Bobs for M protocol rounds.
2. Each party measures the received qubit in the Z basis if the round is classified as a key generation (KG) round, or in the X basis when it is a parameter estimation (PE) round.
3. In the PE step, the parties reveal the outcomes of the PE rounds and estimate the error rate E_X. This information is then used in privacy amplification (PA) to extract a secret conference key. They also reveal a random sample of KG outcomes in order to estimate the error rate affecting their raw keys: $E_{AB_i} = \Pr[Z_A \neq Z_{B_i}]$. Each party now holds a raw key of $n < M$ bits.
4. The parties perform a one-way error correction (EC) procedure where Alice sends sufficient information over the public channel for the Bobs to correct their raw keys and match her key. The amount of information disclosed by Alice is estimated from the error rates E_{AB_i}.
5. In PA the parties map their matching raw keys to a secret conference key by applying the same two-universal hash function, which is randomly picked and publicly broadcast by Alice.

---

[3]This is due to the fact that the N-party GHZ state is an eigenstate of $X^{\otimes N}$ with eigenvalue 1, thus the product of the X outcomes of all the parties must be equal to: $X_A X_{\Pi B} = 1$.

Remark 4.2 *(Preshared key) The parties can know beforehand the classification of each round of the protocol thanks to a preshared secret key they hold. For instance, they could share a key with as many bits as rounds, where the bit value 1 (0) indicates a PE (KG) round. Let us call p_e the probability that a PE round is performed. Since p_e is typically small, the key is mainly composed of zeroes and can thus be highly compressed. In particular, being M the total number of rounds, the parties need a preshared secret key of $M\,h(p_e)$ bits, where $h(x)$ is the binary entropy (c.f. Chap. 2). The length of the preshared key must be subtracted from the final secret key length in order to quantify the amount of fresh secret bits produced by the protocol. However, in the asymptotic regime ($M \to \infty$), the penalty introduced by the preshared key on the asymptotic conference key rate is given by $h(p_e)$ and is negligible, e.g. by choosing $p_e \sim 1/M$.*

In order to know the length of the secret key that can be extracted by PA, the parties need to quantify the smooth min-entropy of Alice's raw key Z_A^n conditioned on Eve's information E (Sect. 2.10), where we emphasized that the raw keys are obtained from Z-basis measurements.

In [13] the min-entropy $H_{\min}^\varepsilon(Z_A^n|E)$ is bounded as a function of E_X, which is computed in the PE step. This is possible thanks to the *uncertainty relation for smooth entropies* [24]. The uncertainty relation states that, given the state $\rho_{AB_1...B_{N-1}E}^n$ of the n rounds yielding the raw keys and assuming that Alice measures her n qubits in either the Z or X basis, it holds:

$$H_{\min}^\varepsilon(Z_A^n|E) \geq q - H_{\max}^\varepsilon(X_A^n|B_1 \ldots B_{N-1}), \tag{4.3}$$

where X_A^n represents the outcomes Alice would obtain had she measured the n KG rounds in the X basis. The term q accounts for the incompatibility of the two measurements of Alice (see [24] for a formal definition). In our case, since Alice measures each qubit either in the Z or X basis, it reads: $q = n$ for n KG rounds.

Thanks to the data-processing inequality (2.66), we can lower bound the r.h.s. of (4.3). Specifically, we assume that every Bob measures his qubit in the X basis in each of the n KG rounds. Thus each Bob B_i obtains a string of X outcomes denoted $X_{B_i}^n$. We then multiply element by element the strings of X outcomes of every Bob and obtain $X_{\Pi B}^n = \prod_{i=1}^{N-1} X_{B_i}^n$. The data-processing inequality states that the uncertainty of the Bobs about X_A^n, quantified by $H_{\max}^\varepsilon(X_A^n|B_1 \ldots B_{N-1})$, can only increase if they process their quantum side information in the way we described. This leads to:

$$H_{\min}^\varepsilon(Z_A^n|E) \geq n - H_{\max}^\varepsilon(X_A^n|X_{\Pi B}^n). \tag{4.4}$$

Finally we remark that the max-entropy in (4.4) can always be upper-bounded by (n times) the binary entropy of the error rate E_X affecting the strings X_A^n and $X_{\Pi B}^n$, with a correction $\Delta(n, \varepsilon)$ due to statistical fluctuations [25]:

$$H_{\max}^\varepsilon(X_A^n|X_{\Pi B}^n) \leq n\,h(E_X + \Delta(n, \varepsilon)). \tag{4.5}$$

One can interpret the inequality (4.5) as the finite version of the equality (3.25) linking the conditional von Neumann entropy of two random variables to their error probability. By combining (4.4) and (4.5) we obtain:

$$H_{\min}^{\varepsilon}(Z_A^n|E) \geq n(1 - h(E_X + \Delta(n, \varepsilon))). \tag{4.6}$$

An important aspect in any CKA protocol is the information leakage during EC. In the N-party BB84 protocol we require every Bob to correct his raw key to match Alice's. By employing one-way EC, Alice needs to publicly broadcast enough information such that even the Bob with the largest amount of errors can correct his key. Since the information leak_i she would send to each B_i only depends on the estimated Z-basis error rate E_{AB_i} but otherwise it's independent of B_i, by broadcasting $\max_i \text{leak}_i$ we ensure that every Bob will be able to correct his raw key. In other words, the leakage of one-way EC in a multipartite QKD protocol is equivalent to that of a bipartite protocol performed with the worst-case Bob.

The asymptotic secret key rate of the N-partite BB84 protocol can be heuristically inferred starting from the secret key length of a bipartite QKD protocol in (3.27). We use Eq. (4.6) to bound the min-entropy term, while we replace the leakage term with $\max_i \text{leak}_i$ according to the argument above. Analogously to Sect. 3.3.3, we estimate the minimum leakage relative to B_i as $\text{leak}_i \approx H_{\max}^{\varepsilon'}(R_A^n|R_{B_i}^n)$ and bound the max-entropy with a version of the bound (4.5) where the relevant error rate is the Z-basis error rate E_{AB_i}. Finally, by taking the asymptotic limit of infinitely many rounds we remove all the corrections due to statistical fluctuations and obtain[4]:

$$r_{N\text{-BB84}} = 1 - h(E_X) - \max_{1 \leq i \leq N-1} h(E_{AB_i}). \tag{4.7}$$

Notably, the resulting key rate reads exactly like the BB84 key rate in (3.26), except for a maximization on the QBERs in the Z basis and a more general definition of E_X.

4.2 Security of CKA

In this Section we describe the functioning of a generic CKA protocol and subsequently prove its security. The content of this Section is mainly drawn from [9, 13].

---

[4]A formal derivation of (4.7) can be found in [13].

4.2.1 General CKA Protocol

The setup in which CKA takes place is a straightforward generalization of that assumed for QKD (c.f. Chap. 3). Every party is linked to a quantum source (or to Alice) by quantum channels. Both the source and the quantum channels may be under Eve's full control. The parties can also communicate over authenticated classical public channels (e.g. phone calls) which may be wiretapped by Eve (see Fig. 4.1).

Alice and $N - 1$ Bobs run a CKA protocol whose goal is to output a set of identical keys $(s_A, s_{B_1}, \ldots, s_{B_{N-1}})$ for Alice and the Bobs, completely unknown to Eve. The protocol could also abort and output the symbol: $s_A = s_{B_1} = \cdots = s_{B_{N-1}} = \perp$.

The steps of a generic N-party CKA protocol read as follows.

1. A source distributes a multipartite entangled state to the N parties for M rounds. From a security perspective, the state of the quantum signals over the M rounds is unknown and given by $\rho_{AB_1 \ldots B_{N-1}}^M$. Eve holds its purification.
2. The parties perform local measurements on each received signal and collect the outcomes. For each measurement, they can randomly choose among certain measurement settings according to the protocol's specifications. Typically, one setting is chosen with higher probability and is used for key generation, while the other(s) form the test rounds. A short preshared key (see Remark 4.2) can indicate to each party what type of measurement to perform in each round. Otherwise, the parties

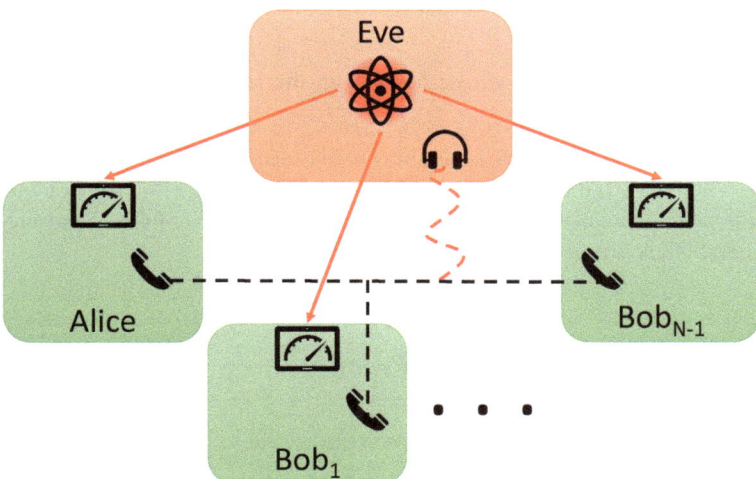

Fig. 4.1 Setup of a CKA protocol with an independent quantum source. Eve may be in control of the source and distribute entangled quantum signals to the parties over quantum channels. Each party locally measures the incoming signal and records the classical output. After the transmission of quantum signals is over, the parties communicate via classical public channels to perform error correction and privacy amplification

can implement a sifting step and select the rounds where they all performed the same type of measurement.

3. For PE the parties reveal the settings and the outcomes of the test rounds, as well as the outcomes of a random sample of key-generation rounds. This information is used to estimate the noise in the quantum channel (Eve's knowledge) and the correlations of their key bits. If the noise is above a certain threshold, the protocol aborts. After this step, Alice and the Bobs hold a string of $n < M$ key-generation outcomes forming their raw key, denoted R_A^n and $R_{B_i}^n$, respectively.

4. In the EC step, each Bob B_i corrects his raw key to match Alice's by computing a guess $\hat{R}_{A_i}^n$ of Alice's raw key. In doing so, the parties reveal leak$_{EC}$ bits of information over the public channel. In order to verify if EC was successful, Alice computes a hash h_A (bitstring) of length $\lceil \log((N-1)/\varepsilon_{EC}) \rceil$ from her raw key R_A^n by applying a randomly-picked two-universal hash function (Definition 2.11). She publicly announces the hash function and h_A. Each Bob uses Alice's hash function to compute the hash h_{B_i} from his guess $\hat{R}_{A_i}^n$. If $h_A \neq h_{B_i}$ for at least one Bob, the protocol aborts. The total amount of information about Alice's raw key R_A^n revealed during EC is thus given by: leak$_{EC}$ + $\lceil \log((N-1)/\varepsilon_{EC}) \rceil \leq$ leak$_{EC}$ + $\log(2(N-1)/\varepsilon_{EC})$.

5. In PA Alice randomly picks another two-universal hash function and broadcasts it. Alice and all the Bobs apply the two-universal hash function on their error-corrected keys and obtain secret conference keys s_A and s_{B_i} (for $i = 1, \ldots, N-1$) of length ℓ. The length ℓ is chosen such that:

$$\ell \leq H_{\min}^{\varepsilon}(R_A^n | E) - \text{leak}_{EC} - \log \frac{2(N-1)}{\varepsilon_{EC}} - 2 \log \frac{1}{2\,\varepsilon_{PA}}, \qquad (4.8)$$

for some $\varepsilon, \varepsilon_{EC}, \varepsilon_{PA} > 0$ which depend on the required level of security (see Sect. 4.2.2).

The crucial task of every CKA protocol is to estimate the smooth min-entropy term in (4.8) with the PE data, as we showed for the multipartite BB84 protocol with (4.6).

In the next Subsection we rigorously define the security of CKA and prove that the general CKA protocol described above is secure.

4.2.2 Security Definition and Proof

Definition 4.1 (*Correctness*) A CKA protocol is said to be ε_{cor}-correct if:

$$\Pr[\cup_{i=1}^{N-1} s_A \neq s_{B_i}] \leq \varepsilon_{cor}. \qquad (4.9)$$

Definition 4.2 (*Secrecy*) A CKA protocol is said to be ε_{sec}-secret if, for Ω being the event that the protocol does not abort, the following inequality holds:

$$\Pr[\Omega]\, T\left(\rho_{S_A E_{\text{tot}}|\Omega}, \omega_{S_A} \otimes \rho_{E_{\text{tot}}|\Omega}\right) \le \varepsilon_{\text{sec}}, \tag{4.10}$$

where $\rho_{S_A E_{\text{tot}}|\Omega}$ is the state that describes the correlation between Alice's final secret key S_A and the total information available to Eve E_{tot} given that the protocol did not abort, while $\omega_{S_A} = \frac{1}{|S|}\sum_{s\in S}|s\rangle\langle s|$ is the maximally mixed state over all the possible realizations of Alice's key and $T(\cdot,\cdot)$ is the trace distance (Definition 2.13).

Definition 4.3 (*Security*) A CKA protocol is said to be ε_{tot}-secure if it is ε_{cor}-correct and ε_{sec}-secret, with $\varepsilon_{\text{tot}} \ge \varepsilon_{\text{cor}} + \varepsilon_{\text{sec}}$.

The correctness definition implies that the protocol always outputs a set of identical keys for all participating parties ($s_A = s_{B_1} = \cdots = s_{B_{N-1}}$), except for probability at most ε_{cor}. The secrecy and security statements are the same used for QKD protocols (c.f. Chap. 3). An ε_{tot}-secure CKA protocol is indistinguishable from an ideal protocol, i.e. one that outputs a set of identical keys unknown to Eve or aborts, except for a probability at most ε_{tot}.

Note that a trivial protocol that always aborts is secure according to the above definition. Thus, another important feature of a CKA protocol is its *completeness*, i.e. the existence of an honest implementation of the protocol such that the probability of not aborting is $\Pr[\Omega] \ge 1 - \varepsilon_c$, for some small ε_c.

We also remark that the CKA security definition is *composable*. This means that when a CKA protocol is composed with another cryptographic task, the security of their combination can be inferred based on their individual security proofs and does not require a separate new proof.

Lemma 4.1 (Security of CKA) *The general CKA protocol of Sect. 4.2.1 is ε_{tot}-secure, with $\varepsilon_{\text{tot}} \ge \varepsilon_{\text{EC}} + 2\varepsilon + \varepsilon_{\text{PA}}$.*

In order to prove this statement, one first shows that the general CKA protocol described earlier is ε_{EC}-correct. This is guaranteed by the fact that the parties verify the success of EC by computing and comparing hashes of length $\lceil \log((N-1)/\varepsilon_{\text{EC}}) \rceil$. The second step is to show that the protocol is at least $(2\varepsilon + \varepsilon_{\text{PA}})$-secret by employing the Quantum Leftover Hash Lemma (c.f. Lemma 2.1), also used to prove the security of QKD protocols. We provide the full proof of Lemma 4.1 in the Appendix of this Chapter (Sect. 4.4).

4.3 Experimental CKA

The first experimental implementation of a CKA protocol has been recently carried out [26], enabling four parties to establish a secret conference key. We also report the realization of a three-party *anonymous* CKA [27], where the identity of the parties establishing the conference key is kept secret from external observers and from each other (except for the initiator of the protocol).

The CKA experiment in [26] implements the multipartite BB84 protocol [13] described in Sect 4.1, by distributing four-party GHZ states encoded in the polarization of single photons at telecom wavelength. In particular, the qubit computational basis $\{|0\rangle, |1\rangle\}$ is encoded in the horizontal/vertical polarization: $\{|H\rangle, |V\rangle\}$. In each round, the photons prepared in the GHZ state are distributed to the four parties over up to a total of 50 km of optical fibres and are subsequently measured in the horizontal/vertical basis (Z basis) or in the diagonal/antidiagonal basis: $\{|+\rangle, |-\rangle\}$ (X basis), where $|\pm\rangle = (|H\rangle \pm |V\rangle)/\sqrt{2}$. After completing the EC and PA steps, the parties hold an ε_{tot}-secure conference key of $\ell = 1.15 \times 10^6$ bits, with $\varepsilon_{tot} = 1.8 \times 10^{-8}$. The established key is then used to encrypt an image with one-time pad in order to securely share it among the four parties.

The generation of the four-party GHZ state is sketched in Fig. 4.2 and is based on the interference of two entangled photon pairs, generated by type-II spontaneous parametric down conversion (SPDC) in periodically poled KTP crystals (PPKTP) embedded in a polarization-based Sagnac interferometer [28].

Two identical sources of entangled photon pairs, S_1 and S_2 in Fig. 4.2a, are supplied by the same mode-locked laser generating light polarized in a coherent superposition of $|H\rangle$ and $|V\rangle$ (blue lines). A $|V\rangle$ photon entering the source is reflected at the

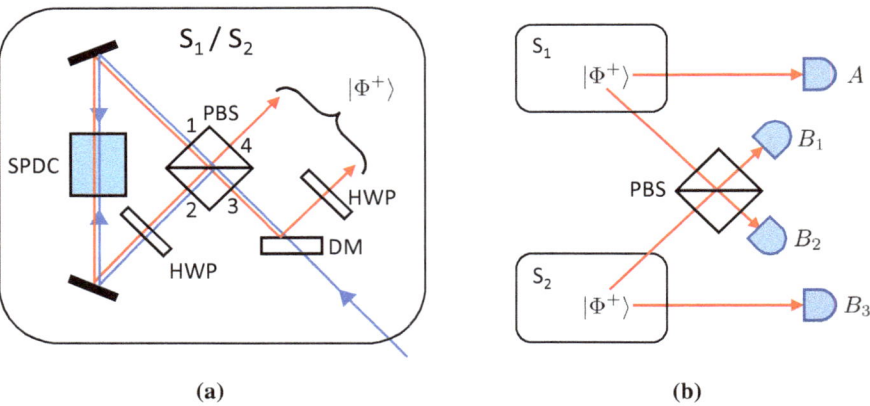

(a)   (b)

Fig. 4.2 Four-party GHZ state generation in the CKA experiment [26] **a** Two identical Bell-pair sources S_1 and S_2 are pumped with the same mode-locked laser (blue line) prepared in a superposition of horizontal and vertical polarization. Vertically polarized photons are reflected at the polarizing beam splitter (PBS), rotated to horizontal polarization by the half-wave plate (HWP) at 45° and undergo type-II spontaneous parametric down-conversion (SPDC) yielding orthogonally polarized photon pairs, which propagate clockwise in the Sagnac loop (red lines). Similarly, horizontally polarized photons are transmitted at the PBS and down-converted to counter-clockwise propagating photon pairs. The interference of the counter-propagating photon pairs at the PBS generates polarization-entangled photons pairs, whose state is described by the Bell state $|\Phi^+\rangle$. **b** One photon from each source interferes in a PBS such that the resulting state of the four photons is the four-party GHZ state, conditioned on the event where each optical path contains one photon, i.e. when Alice and the three Bobs had a click in their detectors. This event occurs with probability 1/2

polarizing beam splitter (PBS)[5] and exits the PBS in path 2. Here, a half-wave plate (HWP) oriented at 45° rotates the photon's polarization to $|H\rangle$ before entering the non-linear crystal. The photon undergoes SPDC in the PPKTP crystal and is converted to two photons of lower energy in path 1 (red line), orthogonally polarized, called signal and idler: $|H_s\rangle_1|V_i\rangle_1$. Similarly, an $|H\rangle$ photon from the laser source gets transmitted at the PBS and is down-converted to signal and idler, $|H_s\rangle|V_i\rangle$, circulating the loop counter-clockwise. Their polarization gets rotated by the HWP in path 2 to: $|V_s\rangle_2|H_i\rangle_2$. At this point, the two counter-propagating photon pairs interfere at the PBS by entering through ports 1 and 2, respectively. The photon pair entering in 1 exits the PBS in the state $|H_s\rangle_3|V_i\rangle_4$, while the pair entering from 2 is mapped to $|V_s\rangle_3|H_i\rangle_4$. The down-converted photons exiting in 3 are separated from the input signal with a dichroic mirror (DM) and rotated by a HWP at 45°. The final state of the photon pair is a coherent superposition of the two considered cases: $|V_s\rangle_3|V_i\rangle_4$ and $|H_s\rangle_3|H_i\rangle_4$, that is the Bell state $|\Phi^+\rangle$:

$$|\Phi^+\rangle = \frac{|H_s\rangle_3|H_i\rangle_4 + |V_s\rangle_3|V_i\rangle_4}{\sqrt{2}}. \tag{4.11}$$

In order to produce a four-party GHZ state, one photon from each Bell pair (4.11) is combined in a PBS before transmitting the photons to Alice and the Bobs (Fig. 4.2b). Indeed, the state of the four photons after interfering at the PBS reads:

$$\frac{1}{2}\left(|HHHH\rangle_{AB_1B_2B_3} + |H\rangle_A|0\rangle_{B_1}|HV\rangle_{B_2}|V\rangle_{B_3} + |V\rangle_A|HV\rangle_{B_1}|0\rangle_{B_2}|H\rangle_{B_3}\right.$$
$$\left. +|VVVV\rangle_{AB_1B_2B_3}\right), \tag{4.12}$$

where $|0\rangle$ indicates the vacuum state and $|HV\rangle_{B_2}$ that there are two photons in path B_2 with orthogonal polarizations. Thus, the parties can post-select the GHZ state $(|HHHH\rangle + |VVVV\rangle)/\sqrt{2}$ by discarding all the events where one of them did not have a detection, which occurs with probability $1/2$.

After interfering one photon per source in the PBS, the four photons are coupled to single-mode fibres of total length up to 50 km and sent to the four parties. Each party measures the incoming photon either in the Z basis ($\{|H\rangle, |V\rangle\}$) for key generation or in the X basis ($\{|+\rangle, |-\rangle\}$) for parameter estimation. Only the events with coincident detections for all parties are retained, since those correspond to the post-selection of a GHZ state.

The measurement setup of each party is depicted in Fig. 4.3. Each party measures the incoming signal in either the X, Y or Z basis by using a quarter-wave plate (QWP), a HWP and a PBS followed by two superconducting nanowire single-photon detectors labelled D_t and D_r. For the Z and X measurements, the QWP is oriented at 0° and the HWP is oriented at 0° and 22.5°, respectively. A click in D_t (D_r) indicates that the photon's polarization has been projected on $|H\rangle$ or $|+\rangle$ ($|V\rangle$ or

---

[5]In a polarizing beam splitter, light polarized horizontally is transmitted while light polarized vertically is reflected.

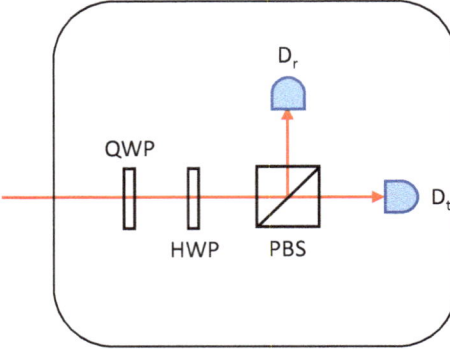

Fig. 4.3 The measurement apparatus of each party in the CKA experiment [26] comprises: a quarter-wave plate (QWP), a half-wave pate (HWP), a polarizing beam splitter (PBS) and two super-conducting nanowire single-photon detectors D_t and D_r collecting the transmitted and reflected photons. The QWP and HWP are motorized in order to implement measurements in the X, Y (for state tomography) and Z basis

$|-\rangle$), depending on the measured basis. The wave plates are motorized and rotate according to the measurement to be performed. Prior to the execution of the protocol, the parties also measure in the Y basis ($|R, L\rangle = (|H\rangle \pm \text{i}|V\rangle)/\sqrt{2}$) to perform state tomography on the post-selected GHZ state. This is achieved with the QWP oriented at $45°$ and the HWP at $0°$.

Once the distribution of quantum states is completed, the parties perform one-way EC from Alice to the Bobs, with a low-density parity-check code (LDPC) adapted to the multipartite scenario. Here, Alice discloses her parity check bits over the public channel in order for the Bobs to compute a guess of her raw key. The parity check matrices are pre-compiled and provided by the Digital Video Broadcasting (DVB-S2) standard [29]. The code rate is chosen such that even the Bob with the largest bit error rate, $\max_i E_{AB_i}$, can successfully correct his raw key (see discussion in Sect. 4.1).

Finally the parties perform privacy amplification by applying an appropriate Toeplitz matrix —a two-universal hash function [30]— randomly picked by Alice on their error-corrected raw keys, obtaining a secret conference key.

The anonymous CKA experiment [27] generates four-photon GHZ states in a similar manner. The only difference is that in [27] the down-converted photon pairs are directly produced as Bell pairs and do not necessitate to be embedded in a Sagnac loop. This can be achieved by using beta-Barium Borate (BBO) crystals for the SPDC process, instead of PPKTP crystals as in [26]. Note that BBO crystals typically yield a lower number of entangled photon pairs compared to PPKTP crystals embedded in a Sagnac loop, when using the same pump power.

Appendix

In this Appendix we prove the security of the general CKA protocol described in Sect. 4.2.1, as stated in Lemma 4.1. A similar proof can be found in [13].

4.4 Finite-key Security of CKA

Proof We first show that the CKA scheme of Sect. 4.2.1 is ε_{EC}-correct. Recall that Alice and the Bobs check if the output of the EC is successful. They do so by comparing the hashes h_A and h_{B_i} (for $i = 1, \ldots, N-1$) obtained by applying the same two-universal hash function on Alice's raw key R_A^n and on the Bobs' guesses $R_{A_i}^{\hat{n}}$, respectively. Each hash is $\lceil \log((N-1)/\varepsilon_{EC}) \rceil$ bits long.

According to Definition 2.11, the probability that two b-bit outputs of a randomly-picked hash function coincide, given that the inputs are different, is given by: 2^{-b}. If there exists a hash h_{B_i} such that $h_{B_i} \neq h_A$, the protocol aborts and outputs the trivial keys $s_A = s_{B_1} = \cdots = s_{B_{N-1}} = \perp$. This implies that the following event has probability zero: $\Pr[s_A \neq s_{B_i}, h_A \neq h_{B_i}] = 0$ and thus $\Pr[s_A \neq s_{B_i}] = \Pr[s_A \neq s_{B_i}, h_A = h_{B_i}]$. Then the CKA is proved to be ε_{EC}-correct, according to Definition 4.1, by the following chain of inequalities:

$$\Pr[\cup_{i=1}^{N-1} s_A \neq s_{B_i}] = \Pr[\cup_{i=1}^{N-1}(s_A \neq s_{B_i}, h_A = h_{B_i})]$$

$$\leq \sum_{i=1}^{N-1} \Pr[s_A \neq s_{B_i}, h_A = h_{B_i}]$$

$$\leq \sum_{i=1}^{N-1} \Pr[h_A = h_{B_i}, R_A^n \neq R_{A_i}^{\hat{n}}]$$

$$\leq \sum_{i=1}^{N-1} \Pr[h_A = h_{B_i} | R_A^n \neq R_{A_i}^{\hat{n}}]$$

$$\leq \sum_{i=1}^{N-1} 2^{-\lceil \log\left(\frac{N-1}{\varepsilon_{EC}}\right) \rceil}$$

$$\leq (N-1)\frac{\varepsilon_{EC}}{N-1}$$

$$\leq \varepsilon_{EC}. \tag{4.13}$$

In the above derivation the first inequality is due to the union bound, the second holds because the final keys s_A and s_{B_i} are obtained from R_A^n and $R_{A_i}^{\hat{n}}$ with another two-universal hash function, and the fourth inequality follows from Definition 2.11.

The CKA secrecy is proved by the Quantum Leftover Hash Lemma [31, 32] exactly like in QKD, since it only concerns the secrecy of Alice's key. The following upper bound holds [31, 32]:

$$\frac{1}{2}\left\|\rho_{S_A E_{\mathrm{tot}}|\Omega} - \omega_{S_A} \otimes \rho_{E_{\mathrm{tot}}|\Omega}\right\| \leq 2\varepsilon + \frac{1}{2}\sqrt{2^{\ell - H_{\min}^{\varepsilon}(R_A^n | CE)}}, \tag{4.14}$$

where ℓ is the length of Alice's key after PA and where we emphasize E_{tot} being the total information available to Eve. This comprises her purifying system E, the classical communication C occurred during EC and the knowledge F of the hash function used in PA: $E_{\mathrm{tot}} = FCE$.

We now employ the following chain-rule for the min-entropy [25]:

$$H_{\min}^{\varepsilon}(R_A^n | CE) \geq H_{\min}^{\varepsilon}(R_A^n | E) - \log|C|$$
$$= H_{\min}^{\varepsilon}(R_A^n | E) - \mathrm{leak}_{\mathrm{EC}} - \log\frac{2(N-1)}{\varepsilon_{\mathrm{EC}}}, \tag{4.15}$$

where $\log|C|$ quantifies all the information revealed during EC and is given by $\mathrm{leak}_{\mathrm{EC}} + \log(2(N-1)/\varepsilon_{\mathrm{EC}})$ (see the CKA description).

By inserting Eq. (4.15) into (4.14) we obtain the following chain of inequalities:

$$\frac{1}{2}\left\|\rho_{S_A E_{\mathrm{tot}}|\Omega} - \omega_{S_A} \otimes \rho_{E_{\mathrm{tot}}|\Omega}\right\| \leq 2\varepsilon + \frac{1}{2}\sqrt{2^{\ell - (H_{\min}^{\varepsilon}(R_A^n | E) - \mathrm{leak}_{\mathrm{EC}} - \log(2(N-1)/\varepsilon_{\mathrm{EC}}))}}$$
$$\leq 2\varepsilon + \frac{1}{2}\sqrt{2^{\log(2\,\varepsilon_{\mathrm{PA}})^2}}$$
$$= 2\varepsilon + \varepsilon_{\mathrm{PA}}, \tag{4.16}$$

where we used the key length expression (4.8) in the second inequality. We have thus proven that the protocol is $\varepsilon_{\mathrm{sec}}$-secret (Definition 4.2), with $\varepsilon_{\mathrm{sec}} \geq 2\varepsilon + \varepsilon_{\mathrm{PA}}$. By combining this with the correctness proof (4.13), we have shown that the protocol is $\varepsilon_{\mathrm{tot}}$-secure, with $\varepsilon_{\mathrm{tot}} \geq 2\varepsilon + \varepsilon_{\mathrm{PA}} + \varepsilon_{\mathrm{EC}}$. This concludes the proof. $\square$

References

1. Epping, M., Kampermann, H., & Bruß, D. (2016a). Large-scale quantum networks based on graphs. *New Journal of Physics, 18*(5), 053036.
2. Epping, M., Kampermann, H., & Bruß, D. (2016b). Robust entanglement distribution via quantum network coding. *New Journal of Physics, 18*(10), 103052.
3. Pirker, A., Wallnöfer, J., & Dür, W. (2018). Modular architectures for quantum networks. *New Journal of Physics, 20*(5), 053054.
4. Hahn, F., Pappa, A., & Eisert, J. (2019). Quantum network routing and local complementation. *npj Quantum Information, 5*(1), 76.

5. Krutyanskiy, V., Meraner, M., Schupp, J., Krcmarsky, V., Hainzer, H., & Lanyon, B. P. (2019). Light-matter entanglement over 50 km of optical fibre. *npj Quantum Information, 5*(1), 72.
6. Tchebotareva, A., Hermans, S. L. N., Humphreys, P. C., Voigt, D., Harmsma, P. J., Cheng, L. K., et al. (2019). Entanglement between a diamond spin qubit and a photonic time-bin qubit at telecom wavelength. *Physical Review Letters, 123*, 063601.
7. Kimble, H. J. (2008). The quantum internet. *Nature, 453*(7198), 1023–1030.
8. Wehner, S., Elkouss, D., & Hanson, R. (2018). Quantum internet: A vision for the road ahead. *Science, 362*(6412).
9. Murta, G., Grasselli, F., Kampermann, H., & Bruß, D. (2020). Quantum conference key agreement: A review. arXiv:quant-ph/2003.10186.
10. Wu, Y., Zhou, J., Gong, X., Guo, Y., Zhang, Z.-M., & He, G. (2016). Continuous-variable measurement-device-independent multipartite quantum communication. *Physical Review A, 93*, 022325.
11. Epping, M., Kampermann, H., Macchiavello, C., & Bruß, D. (2017). Multi-partite entanglement can speed up quantum key distribution in networks. *New Journal of Physics, 19*(9), 093012.
12. Zhang, Z., Shi, R., & Guo, Y. (2018). Multipartite continuous variable quantum conferencing network with entanglement in the middle. *Applied Sciences, 8*(8).
13. Grasselli, F., Kampermann, H., & Bruß, D. (2018). Finite-key effects in multipartite quantum key distribution protocols. *New Journal of Physics, 20*(11), 113014.
14. Grasselli, F., Kampermann, H., & Bruß, D. (2019). Conference key agreement with single-photon interference. *New Journal of Physics, 21*(12), 123002.
15. Ottaviani, C., Lupo, C., R. L., & Pirandola, S. (2019). Modular network for high-rate quantum conferencing. *Communications Physics, 2*(118).
16. Ribeiro, J., Murta, G., & Wehner, S. (2019). Reply to "comment on 'fully device-independent conference key agreement' ". *Phys. Rev. A, 100*, 026302.
17. Pirandola, S. (2019). End-to-end capacities of a quantum communication network. *Communications Physics, 2*(1), 51.
18. Das, S., Bäuml, S., Winczewski, M., & Horodecki, K. (2019). Universal limitations on quantum key distribution over a network. arXiv:quant-ph/1912.03646.
19. Pirandola, S. (2019). General upper bounds for distributing conferencing keys in arbitrary quantum networks. arXiv:quant-ph/1912.11355.
20. Takeoka, M., Kaur, E., Roga, W., and Wilde, M. M. (2019). Multipartite entanglement and secret key distribution in quantum networks. arXiv:quant-ph/1912.10658.
21. Bennett, C. H. and Brassard, G. (1984). Quantum cryptography: Public key distribution and coin tossing. In *Proceedings of IEEE International Conference on Computers, Systems and Signal Processing*, pages 175 – 179.
22. Matsumoto, R. (2007). Multiparty quantum-key-distribution protocol without use of entanglement. *Phys. Rev. A, 76*, 062316.
23. Bruß, D. (1998). Optimal eavesdropping in quantum cryptography with six states. *Phys. Rev. Lett., 81*, 3018–3021.
24. Tomamichel, M., & Renner, R. (2011). Uncertainty relation for smooth entropies. *Phys. Rev. Lett., 106*, 110506.
25. Tomamichel, M., Lim, C. C. W., Gisin, N., & Renner, R. (2012). Tight finite-key analysis for quantum cryptography. *Nature Communications, 3*(1), 634.
26. Proietti, M., Ho, J., Grasselli, F., Barrow, P., Malik, M., and Fedrizzi, A. (2020). Experimental quantum conference key agreement. arXiv:quant-ph/2002.01491.
27. Hahn, F., de Jong, J., Thalacker, C., Demirel, B., Barz, S., and Pappa, A. (2020). Anonymous conference key agreement in quantum networks. arXiv:quant-ph/2007.07995.
28. Fedrizzi, A., Herbst, T., Poppe, A., Jennewein, T., & Zeilinger, A. (2007). A wavelength-tunable fiber-coupled source of narrowband entangled photons. *Opt. Express, 15*(23), 15377–15386.
29. Morello, A., & Mignone, V. (2006). Dvb-s2: The second generation standard for satellite broad-band services. *Proceedings of the IEEE, 94*(1), 210–227.

30. Hayashi, M. (2011). Exponential decreasing rate of leaked information in universal random privacy amplification. *IEEE Transactions on Information Theory, 57*(6), 3989–4001.
31. Renner, R. (2008). Security of quantum key distribution. *International Journal of Quantum Information, 06*(01), 1–127.
32. Tomamichel, M., Schaffner, C., Smith, A., & Renner, R. (2011). Leftover hashing against quantum side information. *IEEE Transactions on Information Theory, 57*(8), 5524–5535.

Chapter 5
Quantum Key Distribution with Imperfect Devices

Abstract In this Chapter we outline some of the experimental flaws that allow a potential eavesdropper to successfully breach the security of a QKD protocol. We then focus on the solutions to such problems, which are given by a combination of theoretical advances and clever experimental design. In particular, in Sects. 5.1 and 5.2 we discuss how security can be proven when the BB84 protocol is implemented with weak coherent pulses instead of single-photon sources, with a particular mention to the decoy-state method. We then introduce measurement-device-independent (MDI) QKD in Sect. 5.3 and consider a practical implementation of it in Sect. 5.4.

The information-theoretic security, which is in principle promised by QKD, is undermined by the difficulty of ensuring that the assumptions on its implementation (quantum sources and/or measurement devices) are met in practice.

5.1 BB84 with Weak Coherent Pulses

The majority of the sources used in QKD experiments are highly attenuated lasers producing weak coherent pulses (WCPs). WCPs are well represented by coherent states:

$$|\alpha\rangle = e^{\frac{-|\alpha|^2}{2}} \sum_{n=0}^{\infty} \frac{\alpha^n}{\sqrt{n!}} |n\rangle,\tag{5.1}$$

where $|n\rangle$ is called a *Fock state* and represents n identical photons, while $|\alpha|^2$ is the intensity of the pulse and represents the average number of photons in the pulse. Indeed, the probability of finding n photons in the coherent state (5.1) follows a Poisson distribution and is given by:

$$\Pr(n) = |\langle n|\alpha\rangle|^2 = e^{-|\alpha|^2} \frac{|\alpha|^{2n}}{n!}.\tag{5.2}$$

© The Author(s), under exclusive license to Springer Nature Switzerland AG 2021
F. Grasselli, *Quantum Cryptography*, Quantum Science and Technology,
https://doi.org/10.1007/978-3-030-64360-7_5

Clearly the number of photons in a WCP is not well defined, opposed to the assumption made in Chap. 3 of Alice sending a single photon to Bob in each round of the protocol. In particular, there is a non-zero probability that Alice sends a multiphoton signal to Bob in one protocol round:

$$p_{\text{multi}} = 1 - \Pr(0) - \Pr(1) = 1 - e^{-|\alpha|^2} - |\alpha|^2 \, e^{-|\alpha|^2} > 0. \tag{5.3}$$

This allows Eve to perform new eavesdropping attacks which severely compromise security, like the photon number splitting attack (PNS) [1, 2].

In a PNS attack, Eve first replaces the lossy channel linking Alice and Bob with a lossless channel. She then performs quantum non-demolition (QND) measurements[1] on the pulses sent by Alice that project them onto subspaces characterized by a fixed photon number, without modifying the pulses' polarization. If the pulse contains just one photon, she blocks it with the same probability of having a loss in the original lossy channel. If she observes multiple photons, she deterministically splits one photon off the signal and stores it in her quantum memory, while sending the remaining photons to Bob. In this way, she has a copy of the photon(s) received by Bob without being noticed. After the parties reveal the bases used in every round, Eve measures the photons in her quantum memory accordingly and learns the key.

From the above example, we learn that only the single-photon signals emitted by Alice are still secure. The security proof by Gottesman-Lo-Lütkenhaus-Preskill (GLLP) [3] states that a BB84 protocol implemented with WCPs is still secure, provided that one extracts the key only from single-photon signals. The resulting asymptotic secret key rate, for an asymmetric BB84 protocol where the Z basis is used for key generation and the X basis for PE, reads [4, 5]:

$$r_{\text{GLLP}} = p_Z^2 \left[Q_Z^0 + Q_Z^1 (1 - h(e_X^1)) - Q_Z h(E_Z) \right], \tag{5.4}$$

where p_Z is the probability that Alice (Bob) chooses the Z basis (asymptotically it can be chosen $p_Z \to 1$). In the GLLP rate (5.4), we recognize the contribution coming from the estimation of Eve's uncertainty from single-photon signals $Q_Z^0 + Q_Z^1 (1 - h(e_X^1))$ from which we subtract the information leaked during error correction (EC) $Q_Z h(E_Z)$, similarly to the asymptotic BB84 rate in (3.26). In particular, Q_Z^n ($n = 0, 1$) is the probability that Alice sent n photons in the Z basis and Bob had a detection event, while Q_Z is the *gain* in the Z basis, i.e. the probability that Bob had a detection given that Alice sent a WCP in that basis. Analogous quantities are defined for the X basis. We have that:

$$Q_{Z(X)} = \sum_{n=0}^{\infty} Q_{Z(X)}^n. \tag{5.5}$$

We note that $Q_Z^0 = Q_X^0 = Q^0$ is independent of the basis, since in this case Bob's detection is caused by dark counts or stray light in his detectors. Hence, the data Bob

---

[1] A QND measurement preserves the physical integrity of the system being measured.

collected in these instances is secure and added to Eve's uncertainty, as there is no way for Eve to know it.

Finally $E_{Z(X)}$ is the QBER in the $Z(X)$ basis given that Bob had a detection, and $e_{Z(X)}^n$ is the error rate in the $Z(X)$ basis given that Alice sent n photons and Bob had a detection. It thus holds:

$$E_{Z(X)} Q_{Z(X)} = \sum_{n=0}^{\infty} Q_{Z(X)}^n e_{Z(X)}^n. \tag{5.6}$$

In a real experiment, the observed quantities are the gains Q_Z, Q_X and the QBERs E_Z, E_X, while p_Z is an input parameter and Q^0, Q_Z^1 and e_X^1 must be estimated. In particular, one can lower bound the achievable key rate (5.4) by upper bounding e_X^1 and by lower bounding Q^0 and Q_Z^1. The decoy-state method [4, 6, 7] provides an excellent way to obtain such bounds, thus guaranteeing high key rates for QKD protocols implemented with WCPs.

Before presenting the decoy-state method in Sect. 5.2, we remark that PNS attacks are not a threat to QKD (or CKA) protocols based on the distribution of entangled states from an untrusted source.

Remark 5.1 [PNS attacks and entanglement] The security proofs of entanglement-based QKD and CKA protocols (c.f. Lemmas 3.1 and 4.1) allow Eve to be in control of the quantum source distributing entangled states to the parties. Indeed, no assumption is made on the state distributed by Eve in each protocol round. In this context, a PNS attack is equivalent to a collective attack where Eve attaches an ancilla photon to the ideal entangled state the parties expect to receive, and keeps the ancillary photon. For instance, in a BB84 protocol Eve would prepare the Bell state $|\Phi^+\rangle$ (c.f. Chap. 3) with two identical photons destined to Alice, and then she keeps one of the two. This means that Eve prepares a three-qubit state and keeps one of the qubits, i.e. she performs a collective attack.

Collective attacks are already accounted for in the security proofs of QKD and CKA protocols and can be readily detected by the parties with parameter estimation. In conclusion, PNS attacks are already included in the security of QKD and CKA based on the distribution of entangled states and do not pose any threat to their security.

5.2 Decoy-State Method

We start by requiring Alice to prepare and send a phase-randomized WCP in each round, whose state is a mixture of Fock states:

$$\rho_\mu = \frac{1}{2\pi} \int_0^{2\pi} d\theta \, |\sqrt{\mu} e^{i\theta}\rangle\langle\sqrt{\mu} e^{i\theta}| = \sum_{n=0}^{\infty} e^{-\mu} \frac{\mu^n}{n!} |n\rangle\langle n|. \tag{5.7}$$

This can be viewed as Alice preparing one of the Fock states $|n\rangle\langle n|$ according to a Poisson distribution like (5.2) with mean photon number μ. Thus the probability of Alice sending exactly n photons in the Z (X) basis and Bob having a detection, $Q_{Z(X)}^n$, is given by:

$$Q_{Z(X)}^n = e^{-\mu} \frac{\mu^n}{n!} Y_{Z(X)}^n, \qquad (5.8)$$

where the n-photon *yield* $Y_{Z(X)}^n$ is the conditional probability that Bob had a detection, given that Alice sent n photons. Again, while the intensity μ of the WCP is an input parameter, the yields are not directly observable.

According to the decoy-state method [4, 6, 7], Alice will intersperse her states ρ_μ used for key generation—called signal states—with decoy states ρ_{μ_i} with the same characteristics of the signal states except for their intensity, which is randomly drawn from a set $\{\mu_i\}_i$ (typically $\mu_i \leq \mu$). This can be achieved with intensity modulators such as variable optical modulators (VOAs). In doing so, the parties observe the gains $Q_{Z(X)}^{\mu_i}$ and QBERs $E_{Z(X)}^{\mu_i}$.

The central idea is that, from Eve's viewpoint, in every round a Fock state $|n\rangle\langle n|$ is picked according to a probability distribution that is unknown to her and sent through the quantum channel. In other words, Eve cannot distinguish a signal state from a decoy state. This means that Eve's action can only depend on the photon number and on the basis (e.g., photon polarization), but not on the probability distribution that generated the photons.

Therefore the yields $Y_{Z(X)}^n$ and the error rates $e_{Z(X)}^n$, which are a reflection of Eve's action on the quantum channel, are independent of the intensity determining the photons' distribution. This fact allows us to derive a set of linear constraints on the yields and error rates, in terms of the observed gains Q_Z, Q_X and QBERs E_Z, E_X. Indeed, by combining (5.8) with (5.5) and (5.6), we obtain:

$$Q_Z^{\mu_i} = \sum_{n=0}^{\infty} e^{-\mu_i} \frac{\mu_i^n}{n!} Y_Z^n \quad, \quad \mu_i \in \{\mu_i\}_i \qquad (5.9)$$

$$Q_X^{\mu_i} = \sum_{n=0}^{\infty} e^{-\mu_i} \frac{\mu_i^n}{n!} Y_X^n \quad, \quad \mu_i \in \{\mu_i\}_i \qquad (5.10)$$

$$E_X^{\mu_i} Q_X^{\mu_i} = \sum_{n=0}^{\infty} e^{-\mu_i} \frac{\mu_i^n}{n!} Y_X^n e_X^n \quad, \quad \mu_i \in \{\mu_i\}_i. \qquad (5.11)$$

Every equality above represents a system of equations determined by different decoy intensities μ_i. The larger the number of decoy intensities, the more constrained are the yields and error rates. By combining the different equations in a system with Gaussian elimination techniques, one can derive bounds on the yields and error rates of interest in terms of the observed gains and QBERs. Importantly, since the employed decoy intensities are typically small (e.g., $\mu_i \sim 0.1$), the higher order terms in each sum can be crudely approximated without heavily affecting the bounds. Moreover, we remark that already two decoy intensities are enough to find good bounds [5, 7,

8] and that recently it was shown that the BB84 protocol can also be implemented with just one decoy intensity setting [9].

From the first set of equations (5.9) one derives lower bounds $Y^{0\downarrow}$ and $Y_Z^{1\downarrow}$ which correspond to lower bounds on the quantities Q^0 and Q_Z^1 appearing in the key rate (5.4). From the second set (5.10), one derives bounds on the yields that are then employed in the third set (5.11) to derive the upper bound $e_X^{1\uparrow}$.

Let us briefly sum up the implementation of the asymmetric BB84 protocol with the integration of the decoy-state method. Alice prepares phase-randomized WCPs polarized in the Z (X) basis with probability p_Z $(1 - p_Z)$. Upon choosing the Z basis, she modulates the pulse intensity to μ with probability q to generate a signal state, or to one of the decoy intensities $\{\mu_i\}$ to generate a decoy state with probability $1 - q$. If Alice picks the X basis instead, she only generates decoy states. Bob chooses to measure the incoming pulse in the Z (X) basis with probability p_Z $(1 - p_Z)$.

At the end of the transmission, Alice reveals the intensity setting and the basis she used in every round. Bob instead reveals all the X outcomes to estimate $E_X^{\mu_i}$ and some of the Z outcomes of the signal state to estimate E_Z^μ.

The asymptotic secret key rate of an asymmetric BB84 protocol with decoy states is obtained with the GLLP analysis and reads [5]:

$$r_{\text{decoy}} \geq p_Z^2 q \left[Q^{0\downarrow} + Q_Z^{1\downarrow}(1 - h(e_X^{1\uparrow})) - Q_Z^\mu h(E_Z^\mu) \right], \tag{5.12}$$

where Q_Z^μ and E_Z^μ are the gain and QBER of the signal state, while $Q^{0\downarrow}$ $(Q_Z^{1\downarrow})$ is a lower bound on the probability that Alice sent 0 (1) photon and Bob had a detection event, given that Alice sent a signal state: $Q^0 = e^{-\mu} Y^{0\downarrow}$ and $Q_Z^{1\downarrow} = e^{-\mu} \mu Y_Z^{1\downarrow}$.

Finally, we mention that the key rate could be optimized by using the decoy-state rounds in the Z basis even for key generation [8].

5.3 Introduction to Measurement-device-independent QKD

The security proof of the general QKD protocol presented in Sect. 3.3 is based on the assumption that the measurement devices held by the parties are trusted, while the source of quantum states can be untrusted. Indeed, we assume that Eve distributes uncharacterised quantum states, on which the parties perform characterised measurements[2] (e.g., in the Z or X basis). All the information that Eve can gain on the measurement outcomes comes from her quantum side information E (apart from the information leaked in the classical public channel).

However, measurement detectors can suffer from imperfections causing them to operate differently from their theoretical models used to prove security. Eve could exploit such imperfections to launch powerful eavesdropping attacks [10–12] that

---

[2]Note that specifying the measurement operators of a party effectively fixes the dimension of the quantum system on which they are performed.

go under the name of *detector side channels*. An example is the *detector blinding attack* [11], where Eve first sends bright light to Bob's single-photon detectors to "blind" them and make them operate in linear-mode. This means that his detectors are now unable to detect single photons and produce a click only above a certain intensity threshold. Eve then sends tailored light pulses to Bob which yield a click only when Bob chooses the same basis in which Eve prepared the pulse. Hence Eve knows the outcome of each detection observed by Bob, without introducing noticeable disturbance.

Measurement-device-independent QKD (MDI-QKD) [13, 14] provides a solution which removes all possible detector side channels with a new QKD paradigm. Here, the honest parties send quantum signals to an intermediate relay which applies some measurement and publicly announces the outcome. The founding idea is to remove all trust from the measurement apparatus, which can be operated by Eve, and place it on the sources, held by Alice and Bob. Typically, QKD sources are attenuated lasers which can be easily characterized in a controlled environment such as Alice's and Bob's laboratories. Note that this scenario is opposite to the previous one, where the source was untrusted and the measurement devices were trusted.

Despite the fact that Eve has potentially full control on the relay and on the connecting quantum channels, Alice and Bob can still establish a secret key. This is possible if the measurement outcome publicly announced by the relay, in an honest implementation, is informative for Alice and Bob but is not informative—i.e. it does not reveal information on the key—for anyone else, including Eve.

To make things more concrete, let us consider an idealized MDI-QKD proto-col [15] where Alice and Bob independently encode their bits in the rectilinear or diagonal polarization of single-photon states, represented by the bases $\{|0\rangle, |1\rangle\}$ and $\{|+\rangle, |-\rangle\}$ (with $|\pm\rangle = (|0\rangle + |1\rangle)/\sqrt{2}$), respectively. The quantum signals are then sent to the relay. Here a Bell-state measurement, i.e. a projection on one of the four Bell states $|\psi_{ij}\rangle$ given in (3.15), is applied on the incoming signals and its outcome (i, j) is announced. Upon sifting, Alice and Bob are only left with bits corresponding to rounds in which they used the same basis. If both parties used the rectilinear basis, the outcomes $(i, 0)$ (for $i = 0, 1$) inform them that their bit values coincide, while the outcomes $(i, 1)$ indicate that they encoded opposite bit values. Similarly, if Alice and Bob used the diagonal basis, the outcomes $(0, j)$ indicate they have same bit values while $(1, j)$ indicate opposite bit values. Bob can thus flip his bit according to the measurement outcome and recover Alice's bit.

In simple terms, the outcome of the Bell-state measurement reveals the parity of the parties' bits but not their values. Therefore, it provides useful information only if one of the two bit values is known (i.e. to Alice and Bob), while being useless otherwise.

Of course, Eve could implement any other operation on the incoming pulses but she is still required to announce an outcome of the form (i, j) at every round. Thus, by comparing a fraction of their sifted bits, Alice and Bob can verify the deviation of the actual measurement apparatus from the ideal one and quantify the amount of information gained by Eve.

5.4 Practical Measurement-Device-Independent QKD

The first practical version of an MDI-QKD protocol was introduced by Lo and co-workers in [13] with a fully optical setup. The single-photon pulses are replaced by phase-randomized WCPs in combination with the decoy-state method to guarantee security, while the Bell-state measurement is implemented using linear optics. Unfortunately, with linear optics only two out of four Bell-state projectors can be realized. This, however, does not undermine security as it only introduces some inconclusive measurement outcomes, which reduce the key rate. The protocol's setup is reported in Fig. 5.1.

In every round of the protocol, Alice (Bob) prepares a phase-randomized WCP. Upon randomly selecting the rectilinear (horizontal, vertical) or diagonal ($45°$, $-45°$) polarization basis, Alice (Bob) encodes a random bit in the polarization state of the pulse with a polarization modulator. The amplitude of the pulse is randomly tuned through an amplitude modulator, generating signal or decoy states. The two pulses are then sent to the central relay where they interfere at a 50:50 beam splitter (BS). At each output port of the BS, a polarizing beam splitter (PBS) projects the incoming pulses onto the horizontal (H) or vertical (V) polarization states, which are then detected by the corresponding single-photon detectors (SPDs): D_{C_H}, D_{D_H} (horizontal) and D_{C_V}, D_{D_V} (vertical). The outcome of the detections is publicly announced.

The click of exactly two detectors corresponding to orthogonal polarizations indicates a successful Bell-state measurement. In particular if D_{C_H}, D_{C_V} or D_{D_H}, D_{D_V} clicked, the pulses have been projected on the Bell state $|\psi_{01}\rangle$. If a click occurs in D_{C_H}, D_{D_V} or D_{D_H}, D_{C_V}, the projection is on the Bell state $|\psi_{11}\rangle$. All other detection

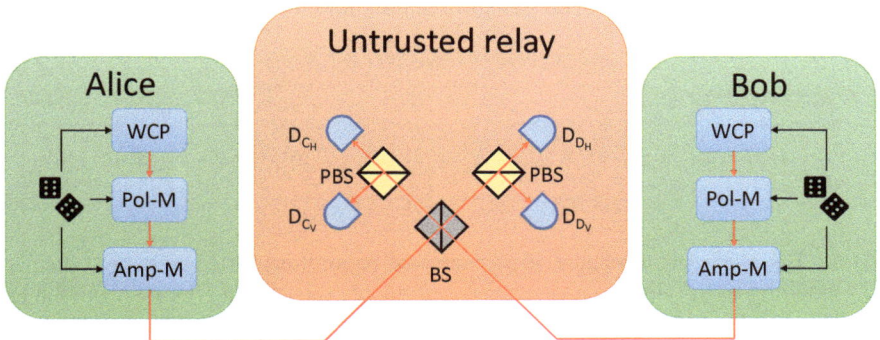

Fig. 5.1 Schematic setup of the MDI-QKD protocol in [13]. Each party prepares a phase-randomized weak coherent pulse (WCP). With a polarization modulator (Pol-M), the party encodes a random bit in the polarization of the pulse. The intensity of the pulse is attenuated with an amplitude modulator (Amp-M) to implement the decoy state method. The parties send their pulses to the central relay, which in principle applies a 50:50 beam splitter (BS) followed by two polarizing beam splitters (PBS) at each output port, which project the incoming pulses on the horizontal or vertical polarization states. The resulting pulses are detected by four single-photon detectors (D_{C_H}, D_{C_V}, D_{D_H}, D_{D_V}). The detection pattern is publicly revealed

events are inconclusive and the corresponding bits get discarded. The parties also discard the bits for which they used different bases. Bob flips all his remaining bits, except for those generated in rounds where he selected the diagonal basis and the pulses were projected on $|\psi_{01}\rangle$.

In order to grasp how the optical setup depicted in Fig. 5.1 corresponds to a Bell-state measurement, we imagine a virtual scenario where the state preparation goes as follows. We assume for simplicity that the parties can prepare single-photon states. Alice (and similarly Bob) prepares an entangled state between a virtual qubit she (he) holds and a single photon polarized either horizontally or vertically:

$$|\Phi_A^+\rangle = \frac{1}{\sqrt{2}}\left[|H\rangle_A|1\rangle_{A_H} + |V\rangle_A|1\rangle_{A_V}\right] = \left[|H\rangle_A a_H^\dagger + |V\rangle_A a_V^\dagger\right]|0\rangle \qquad (5.13)$$

$$|\Phi_B^+\rangle = \frac{1}{\sqrt{2}}\left[|H\rangle_B|1\rangle_{B_H} + |V\rangle_B|1\rangle_{B_V}\right] = \left[|H\rangle_B b_H^\dagger + |V\rangle_B b_V^\dagger\right]|0\rangle. \qquad (5.14)$$

The kets $|H\rangle_A, |V\rangle_A$ define the qubit's computational basis (Z basis) indicating the polarization state the single photon, while the Fock states $|1\rangle_{A_H}, |1\rangle_{A_V}$ describe a single photon polarized horizontally or vertically, and can be expressed in terms of the corresponding creation operators $a_H^\dagger, a_V^\dagger$ acting on the vacuum $|0\rangle$. Analogous definitions hold for Bob's state.

If now Alice (Bob) measures the virtual qubit in the Z or X basis, this is equivalent to Alice (Bob) preparing the single-photon in a random polarization state of the corresponding basis, which is the protocol's state preparation described above. However, since Alice and Bob's measurements commute with the detection at the relay, they can be delayed until the photon detection has occurred.

Therefore, after preparing the entangled states (5.13) and (5.14), the parties send their photons to the relay. The global quantum state before the photons enter the 50:50 BS reads:

$$|\Phi_A^+\rangle \otimes |\Phi_B^+\rangle =$$
$$\frac{1}{2}\left[|HH\rangle_{AB} a_H^\dagger b_H^\dagger + |HV\rangle_{AB} a_H^\dagger b_V^\dagger + |VH\rangle_{AB} a_V^\dagger b_H^\dagger + |VV\rangle_{AB} a_V^\dagger b_V^\dagger\right]|0\rangle.$$
$$(5.15)$$

At the BS, every photon has a 50% chance of being transmitted or being reflected. By labelling $c^\dagger$ ($d^\dagger$) the creation operator of the photons exiting the BS from the left (right) output port (c.f. Fig. 5.1), the unitary action of the BS can be summarized as follows:

$$a^\dagger \mapsto \frac{c^\dagger + d^\dagger}{\sqrt{2}} \qquad (5.16)$$

$$b^\dagger \mapsto \frac{c^\dagger - d^\dagger}{\sqrt{2}}. \qquad (5.17)$$

By inserting the relations (5.16) and (5.17) in the state (5.15) and by using the fact that creation operators relative to different optical paths or different polarization states commute, we obtain the following global state at the exit of the BS:

$$|\Phi_{BS}\rangle =$$
$$\frac{1}{2}\left[|\psi_{01}\rangle_{AB}\left(\frac{|1\rangle_{C_H}|1\rangle_{C_V} - |1\rangle_{D_H}|1\rangle_{D_V}}{\sqrt{2}}\right) - |\psi_{11}\rangle_{AB}\left(\frac{|1\rangle_{C_H}|1\rangle_{D_V} - |1\rangle_{C_V}|1\rangle_{D_H}}{\sqrt{2}}\right)\right.$$
$$\left. +|HH\rangle_{AB}\left(\frac{|2\rangle_{C_H} - |2\rangle_{D_H}}{2}\right) + |VV\rangle_{AB}\left(\frac{|2\rangle_{C_V} - |2\rangle_{D_V}}{2}\right)\right]. \tag{5.18}$$

In the last expression $|\psi_{01}\rangle_{AB}$ and $|\psi_{11}\rangle_{AB}$ are Bell states written in the computational basis $\{|H\rangle, |V\rangle\}$, so for example $|\psi_{01}\rangle_{AB} = (|HV\rangle + |VH\rangle)/\sqrt{2}$, and e.g., $|1\rangle_{C_H}, |2\rangle_{C_H}$ are the Fock states of one and two photons polarized horizontally in the left output port of the BS.

From (5.18) we immediately deduce that if the detectors D_{C_H}, D_{C_V} or D_{D_H}, D_{D_V} click, then the qubits have been projected on the Bell state $|\psi_{01}\rangle_{AB}$. If the clicks occur in D_{C_H}, D_{D_V} or D_{C_V}, D_{D_H}, the qubits are projected on $|\psi_{11}\rangle_{AB}$. Since the qubits indicate the polarization state of the photons, this confirms that the optical setup performs the mentioned Bell state measurement. Moreover, we observe that even in an ideal scenario (single-photons and no losses), this implementation of a Bell state measurement cannot succeed with probability higher than $1/2$, thus reducing the key rate.

Note that the click of only one detector would reveal the polarization of both photons (in absence of losses), hence this event cannot be used for MDI-QKD. The same thing would happen if both detectors D_{C_H}, D_{D_H} or D_{C_V}, D_{D_V} clicked. However, this event cannot happen (c.f. (5.18)) due to the *Hong-Ou-Mandel (HOM) effect*. The HOM effect occurs when two identical photons (like Alice's and Bob's photons when prepared with the same polarization) enter the input ports of a 50:50 BS. Due to the unitary nature of the BS, the two photons always exit the same output port of the BS. If the HOM interference would not occur, the possible detection patterns at the relay would increase, making the Bell state measurement less likely and thus harming the key rate.

Consequently, preparing indistinguishable photons from independent light sources and obtaining good HOM interference is an important requirement for a successful implementation of the described protocol. For this, the authors in [13] also show that such a requirement can be fulfilled with current technology.

The virtual qubit approach also plays a fundamental role in proving the security of the MDI-QKD protocol here presented. Indeed, in the virtual scenario and after the photon detection has occurred, the protocol can be interpreted as an entanglement-based BB84 protocol where Alice and Bob are given a pair of qubits in an entangled state, which ideally is either $|\psi_{01}\rangle$ or $|\psi_{11}\rangle$. The parties then independently measure their qubit in the Z or X basis and compute the QBERs. In this way, one can prove the security of the MDI scheme by relying on the security proof of the BB84 protocol with WCPs [3, 13] (c.f. Sect. 5.1). Note that in this case the secure bits generated by the MDI-QKD protocol are those where both Alice and Bob sent a single photon to the relay. Moreover, a detection event is successful only when exactly two detectors clicked in the combinations described above.

The authors in [13] provide the asymptotic secret key rate achieved by their MDI-QKD protocol. They consider a version of the protocol where the rectilinear basis is used for key generation and the diagonal basis (selected in a small fraction of rounds) is used for estimating Eve's knowledge (PE). The resulting secret key rate reads:

$$r_{\text{MDI}} = Q_{\text{rect}}^{1,1}(1 - h(e_{\text{diag}}^{1,1})) - Q_{\text{rect}}h(E_{\text{rect}}), \tag{5.19}$$

where Q_{rect} and E_{rect} are the gain and QBER of the signal state in the rectilinear basis. That is, Q_{rect} is the probability of a successful detection given that both Alice and Bob sent a signal state in the rectilinear basis. Instead, $Q_{\text{rect}}^{1,1}$ is the probability that both parties sent one photon and the relay had a successful detection, given that they both prepared a signal state in the rectilinear basis. Finally $e_{\text{diag}}^{1,1}$ is the error rate in the diagonal basis given that Alice and Bob sent one photon each and the detection was successful.

As expected, the secret key rate in (5.19) resembles the one in (5.12) of an asymmetric BB84 protocol with decoy states. Similarly to that case, the quantities $Q_{\text{rect}}^{1,1}$ and $e_{\text{diag}}^{1,1}$ can be bounded with the decoy state method presented in the previous Section.

References

1. Huttner, B., Imoto, N., Gisin, N., & Mor, T. (1995). Quantum cryptography with coherent states. *Physical Review A, 51*, 1863–1869.
2. Brassard, G., Lütkenhaus, N., Mor, T., & Sanders, B. C. (2000). Limitations on practical quantum cryptography. *Physical Review Letters, 85*, 1330–1333.
3. Gottesman, D., Lo, H.-K., Lütkenhaus, N., & Preskill, J. (2004). Security of quantum key distribution with imperfect devices. *Quantum Information & Computation, 4*, 325–360.
4. Lo, H.-K., Ma, X., & Chen, K. (2005). Decoy state quantum key distribution. *Physical Review Letters, 94*, 230504.
5. Wei, Z., Wang, W., Zhang, Z., Gao, M., Ma, Z., & Ma, X. (2013). Decoy-state quantum key distribution with biased basis choice. *Scientific Reports, 3*(1), 2453.
6. Hwang, W.-Y. (2003). Quantum key distribution with high loss: Toward global secure communication. *Physical Review Letters, 91*, 057901.
7. Wang, X.-B. (2005). Beating the photon-number-splitting attack in practical quantum cryptography. *Physical Review Letters, 94*, 230503.
8. Lim, C. C. W., Curty, M., Walenta, N., Xu, F., & Zbinden, H. (2014). Concise security bounds for practical decoy-state quantum key distribution. *Physical Review A, 89*, 022307.
9. Rusca, D., Boaron, A., Grünenfelder, F., Martin, A., & Zbinden, H. (2018). Finite-key analysis for the 1-decoy state qkd protocol. *Applied Physics Letters, 112*(17), 171104.
10. Zhao, Y., Fung, C.-H. F., Qi, B., Chen, C., & Lo, H.-K. (2008). Quantum hacking: Experimental demonstration of time-shift attack against practical quantum-key-distribution systems. *Physical Review A, 78*, 042333.

11. Lydersen, L., Wiechers, C., Wittmann, C., Elser, D., Skaar, J., & Makarov, V. (2010). Hacking commercial quantum cryptography systems by tailored bright illumination. *Nature Photonics*, *4*(10), 686–689.
12. Gerhardt, I., Liu, Q., Lamas-Linares, A., Skaar, J., Kurtsiefer, C., & Makarov, V. (2011). Full-field implementation of a perfect eavesdropper on a quantum cryptography system. *Nature Communications*, *2*(1), 349.
13. Lo, H.-K., Curty, M., & Qi, B. (2012). Measurement-device-independent quantum key distribution. *Physical Review Letters*, *108*, 130503.
14. Xu, F., Curty, M., Qi, B., & Lo, H. (2015). Measurement-device-independent quantum cryptography. *IEEE Journal of Selected Topics in Quantum Electronics*, *21*(3), 148–158.
15. Pirandola, S., Andersen, U. L., Banchi, L., Berta, M., Bunandar, D., Colbeck, R., Englund, D., Gehring, T., Lupo, C., Ottaviani, C., Pereira, J., Razavi, M., Shaari, J. S., Tomamichel, M., Usenko, V. C., Vallone, G., Villoresi, P., & Wallden, P. (2019). Advances in quantum cryptography. arXiv:quant-ph/1906.01645.

Chapter 6
Beyond Point-to-Point Quantum Key Distribution

Abstract In this chapter we present the recently-derived theoretical limits on the secret key rate that can be extracted by any point-to-point QKD protocol (Sect. 6.1). Subsequently, we present in detail the (arguably) simplest solution found by researchers to overcome such limitations, which is twin-field (TF) QKD. TF-QKD overcomes the point-to-point private capacity by exploiting single-photon interference in an intermediate untrusted station (Sects. 6.2 and 6.3). This is followed by a detailed investigation of the performance of TF-QKD in realistic conditions, namely finite number of decoy intensity settings and asymmetric channels (Sect. 6.4). Finally, in Sect. 6.5 we generalize the founding idea of TF-QKD to a multiparty scenario by describing a quantum conference key agreement (CKA) based on single-photon interference events.

By definition, point-to-point QKD protocols are implemented with a single quantum channel which directly connects the two parties establishing the key, e.g., the BB84 protocol introduced in Chap. 3.

6.1 Fundamental Limits of Point-to-Point QKD

The secret key rate of any QKD protocol is limited by the losses that inevitably occur in the quantum channel(s) linking the end users. In most QKD implementations, the information is encoded in one of the degrees of freedom of photons. The photons are then transmitted over lossy quantum channels, whose *transmittance* η represents the probability that a photon is successfully transmitted.

For instance, the optical attenuation in standard telecom fibres is about $\gamma = 0.2$ dB km^{-1}, which leads to an overall loss of γL over L kilometres of fibre. The transmittance of an L-kilometre telecom fibre is thus given by: $\eta = 10^{-\gamma L/10}$. This shows that the probability of a photon being transmitted decreases exponentially with the length of the channel, thus severely affecting the key rate.

The exact relation between the key rate and the channel transmittance depends on the protocol. Nevertheless, researchers have recently derived fundamental bounds

F. Grasselli, *Quantum Cryptography*, Quantum Science and Technology,
https://doi.org/10.1007/978-3-030-64360-7_6

[1, 2] on the secret key rate of any *point-to-point* QKD protocol, which only depend on the channel transmittance η. In particular, the secret key rate of any QKD protocol performed over a lossy channel of transmittance η is upper bounded by the Pirandola-Laurenza-Ottaviani-Banchi (PLOB) bound [2]:

$$r_{\text{PLOB}} = -\log(1 - \eta), \tag{6.1}$$

where the logarithm is intended in base 2, as usual. In the high-loss regime ($\eta \ll 1$), we can expand the logarithm in (6.1):

$$r_{\text{PLOB}} \approx 1.44\, \eta, \tag{6.2}$$

and observe that the key rate cannot scale better than linearly with the transmittance of the channel, thus decreasing exponentially with the channel length.

The only way to overcome such severe limitations on the achievable key rate is to employ one or more intermediate nodes in the quantum channel connecting the users. However, this fact alone is not sufficient in general to yield key rates with an improved scaling compared to the PLOB bound.

Consider for instance the MDI-QKD protocol presented in the Chap. 5. Despite featuring an intermediate measuring station that splits the channel between Alice and Bob of transmittance η in two channels of transmittance $\sqrt{\eta}$ each,[1] the key rate does not scale better than the PLOB bound. Indeed, in order to have a successful detection, both photons sent by Alice and Bob need to arrive at the central relay, which occurs with probability $\sqrt{\eta} \cdot \sqrt{\eta} = \eta$. Thus, the gain and hence the key rate cannot scale better than linearly with the transmittance η of the whole channel.

A possible solution is instead represented by quantum repeaters [3, 4], which guarantee a polynomial scaling of the communication efficiency with the distance. However, such devices are still very challenging to implement as they require either quantum memories [4–6] or quantum error correction [7, 8].

Other viable options are evolutions of the original MDI-QKD scheme, like memory-assisted MDI-QKD featuring quantum memories [9, 10] or adaptive MDI-QKD with quantum non-demolition measurements [11]. In both cases, the protocol adapts to the photon losses ensuring that the Bell-state measurement is performed between pulses from Alice and Bob that actually arrived, even combining pulses sent in different rounds. In this way, only one photon per round is required to arrive, thus yielding a key rate proportional to $\sqrt{\eta}$. Despite the square-root improvement in the key rate scaling, the evolved MDI schemes still rely on two-photon interference events and their implementation is far from being practical.

In the next Section we introduce a novel QKD protocol which represents the simplest solution, found so far, to improve the key rate scaling of QKD beyond the PLOB bound thus reaching further distances.

---

[1]Each channel has length $L/2$, if L is the total channel length between Alice and Bob. The transmittance of each channel is thus $e^{-\gamma L/(2\cdot 10)} = \sqrt{\eta}$.

6.2 Twin-Field QKD: Original Protocol

In 2018, Lucamarini et al. [12] proposed a new QKD scheme which is based on the same working principle of MDI-QKD: a central untrusted relay measures the pulses sent by Alice and Bob. The measurement outcome reveals the parity of Alice and Bob's bits, but not their values. However, opposed to MDI-QKD, it is based on *single-photon* interference events. Thanks to this feature, it naturally retains the square-root improvement in the key rate scaling since the successful events are exactly those where only one photon arrived, sent either from Alice or Bob. This removes the necessity of sophisticated systems to adapt the Bell-state measurements to photon losses.

The protocol in [12] takes the name of twin-field (TF) QKD and its original formulation goes as follows. Alice (Bob) generates phase-randomized WCPs by picking a random phase value ρ_a (ρ_b) in the interval $[0, 2\pi)$. The phase interval is split into M phase slices $\Delta_k = 2\pi k/M$ ($k = 0, \ldots, M - 1$) and the selected random phase necessarily falls into one of them: $\Delta_{k(a)}$ ($\Delta_{k(b)}$). Alice (Bob) then encodes a secret bit and a secret basis in another phase φ_a (φ_b) which is added to the phase of the pulse. The pulses are then sent to a central station where they are combined in a 50:50 beam splitter with single-photon detectors at its output ports. After the detection outcome is announced, the parties publicly reveal the slices $\Delta_{k(a)}$, $\Delta_{k(b)}$ and the encoded bases, and keep only the rounds with matching values. Indeed, the optical fields whose random phase falls in the same slice are "twins" and can be used to generate a secret key. The detection outcome combined with the revealed information indicate to Bob whether he needs to flip his bit or not, in order to match it with Alice's.

Since the first TF-QKD protocol has been published, an intense research activity led to several variants of the original scheme [13–17] and to many experimental demonstrations [18–22]. In particular, experimentalists managed to obtain secret key rates surpassing the limit imposed by the PLOB bound, thus proving the improved scaling of TF-QKD.

In the following, we are going to focus on the TF-QKD protocol proposed by Curty et al. [13], which is simpler and arguably better-performing than many other TF-QKD variants. Before moving on, we briefly mention a couple of drawbacks of the original TF scheme in [12].

Firstly, the random phases of a pair of twin fields are not identical and differ by less than $2\pi/M$. This induces an intrinsic QBER that tends to zero for $M \to \infty$. However, the probability of having matching slices scales as $1/M$, thus increasing M leads to more discarded rounds. There exists an optimal value for M that can be determined by appropriately modelling the experimental setup and optimizing the key rate. The authors in [12] obtained an optimal value of $M_{\text{opt}} = 16$. In any case, the use of locally randomized phases by Alice and Bob and the post-selection of the matching ones causes a consistent amount of rounds to be discarded.

The main challenge in implementing the TF-QKD protocol of [12] is controlling the phase drifts of the twin fields. Indeed, the differential phase fluctuation between the two signals sent by Alice and Bob can be quantified as follows:

$$\delta_{ab} = \frac{2\pi}{s} \left(\Delta \nu L + \nu \Delta L \right), \tag{6.3}$$

where s is the speed of light in the fibre, while $\Delta \nu$ is the frequency difference between the users' lasers and ΔL is the difference of path lengths travelled by the two signals. While $\Delta \nu$ can be compensated with phase-locking techniques already used in optical communications, the contribution due to ΔL is a more serious drawback. Indeed, even if the two fibres have nominally the same length, the distance travelled by each signal can fluctuate in time due for instance to thermal expansions of the fibres, resulting in a phase drift at the output of the fibre. This issue could be mitigated with active stabilization techniques.

6.3   Twin-Field QKD Without Phase Post-selection

The TF-QKD protocol introduced in [13] removes the need to post-select matching global phases ρ_a and ρ_b for Alice and Bob, thus sensibly increasing the protocol's performance. It instead relies on a preselected global phase shared by Alice and Bob. We remark that a very similar scheme has been independently developed in [14], but it is equipped with an alternative security proof.

6.3.1   Idealized Protocol

In order to elucidate the protocol's functioning, we start by presenting an idealized version of it, which also has the merit to show where the inspiration came from, namely entanglement-generation protocols in quantum repeaters.

The idealized TF-QKD protocol in [13] is composed of the following steps.

1. Alice (and analogously Bob) prepares an optical signal entangled with a qubit she holds:

$$|\Phi\rangle_{Aa} = \sqrt{q}|0\rangle_A|0\rangle_a + \sqrt{1-q}|1\rangle_A|1\rangle_a \quad 0 \leq q \leq 1, \tag{6.4}$$

   where $|0\rangle_a$ and $|1\rangle_a$ are the Fock states of the photon representing the vacuum and a single-photon state, while $\{|0\rangle_A, |1\rangle_A\}$ is the qubit's computational basis (Z basis).
2. Both parties send their optical pulses to a central relay through optical channels, each with transmittance $\sqrt{\eta}$ (the overall transmittance between Alice and Bob is η).

3. The central relay applies a 50:50 beam splitter to the incoming pulses followed by two threshold detectors D_c and D_d (i.e. unable to distinguish the detection of one or more photons).
4. The relay broadcasts the outcomes k_c and k_d of detector D_c and D_d, where $k_c = 0$ and $k_c = 1$ ($k_d = 0$ and $k_d = 1$) correspond to a no-click and a click event, respectively.
5. With probability p_X Alice (Bob) measures her (his) qubit in the X basis given by $\{|\pm\rangle_{A(B)} = (|0\rangle_{A(B)} \pm |1\rangle_{A(B)})/\sqrt{2}\}$, while with probability $1 - p_X$ she (he) measures the qubit in the Z basis. Upon obtaining the outcome x, where $x = \pm 1$ are the eigenvalues of the X and Z operators, Alice (Bob) records the bit value b_A (b_B) with $(-1)^{b_A} = x$ $((-1)^{b_B} = x)$.
6. The bits b_A and $b_B \oplus k_d$, collected by Alice and Bob in the rounds where they measured in the X basis and where the relay announced $k_c \oplus k_d = 1$ (i.e. only one detector clicked), form their raw keys. The bits collected in the Z-basis rounds where $k_c \oplus k_d = 1$ are instead used for PE. All the other rounds are discarded.

To understand why the protocol enables the parties to distil a secret key, imagine that we choose $1 - q \ll 1$ in the state preparation. This means that both parties prepare their signals strongly unbalanced towards the vacuum. For this reason, in the relevant events where only one detector clicked, the detection is likely to be caused by the sending and arrival of just one photon coming from either Alice or Bob. However, the beam splitter creates a coherent superposition of these two possibilities, implying that either Alice's or Bob's qubit are in state $|1\rangle$, but not both of them. The conditional state of the parties' qubits is thus well approximated by the Bell states: $|\psi_{k_d 1}\rangle_{AB} = (|01\rangle + (-1)^{k_d}|10\rangle)/\sqrt{2}$ ($k_d = 0, 1$). The parties then measure their respective qubit in either the X or Z basis. From here, the protocol can be regarded as an entanglement-based BB84 protocol whose security is proved Chap. 3. Any deviation from the described picture can be detected by computing appropriate error rates.

The states $|\psi_{k_d 1}\rangle$ prompt us to define the following error rates in the X and Z basis:

$$E_X = p_{XX}[b_A \neq b_B \oplus k_d | k_c \oplus k_d = 1] \tag{6.5}$$
$$E_Z = p_{ZZ}[b_A = b_B | k_c \oplus k_d = 1], \tag{6.6}$$

where $p_{XX}[\Omega]$ ($p_{ZZ}[\Omega]$) is the probability that the event Ω occurred given that both Alice and Bob measured in the X (Z) basis. The two error rates are zero if the parties share the Bell state $|\psi_{01}\rangle$ or $|\psi_{11}\rangle$.

According to the above explanation, ideally the relevant detections are caused by the sending and arrival of just one photon. This shows that the protocol is based on *single-photon interference* events, thus producing a key rate that scales with $\sqrt{\eta}$ (the transmittance of one of the two channels) as the original TF-QKD scheme.

We can support this statement with more analytical grounds, by first computing the conditional state of the parties' qubits, given that only detector D_c ($k_d = 0$) or only detector D_d ($k_d = 1$) clicked:

$$\rho_{AB}^{k_d} = \frac{p_1}{p_{\text{click}}} \left[\frac{q}{q + (1-q)(1-\sqrt{\eta})} |\psi_{k_d 1}\rangle\langle\psi_{k_d 1}|_{AB} \right.$$
$$\left. + \frac{(1-q)(1-\sqrt{\eta})}{q + (1-q)(1-\sqrt{\eta})} |11\rangle\langle11|_{AB} \right] + \frac{p_2}{p_{\text{click}}} |11\rangle\langle11|_{AB}, \qquad (6.7)$$

where $p_{\text{click}} = p_1 + p_2$ is the probability that only detector D_c (D_d) clicked, while p_1 (p_2) corresponds to the event where the detection was caused by a single-photon (two-photon) pulse:

$$p_1 = \sqrt{\eta}(1-q)q + (1-q)^2 \sqrt{\eta}(1-\sqrt{\eta}) \qquad (6.8)$$

$$p_2 = \frac{1}{2}(1-q)^2 \eta. \qquad (6.9)$$

In (6.7) we recognize the contribution due to the Bell states $|\psi_{k_d 1}\rangle$, however we also have other spurious contributions which lead to intrinsic errors in the protocol. The resulting error rates, for the state (6.7), read:

$$2E_X = E_Z = \frac{p_1}{p_{\text{click}}} \frac{(1-q)(1-\sqrt{\eta})}{q + (1-q)(1-\sqrt{\eta})} + \frac{p_2}{p_{\text{click}}}. \qquad (6.10)$$

The asymptotic secret key rate of the described protocol is simply given by the BB84 protocol key rate (3.26), rescaled by the probability $2p_{\text{click}}$ that a successful detection occurred. We thus get:

$$r_{\text{idealTF}} = 2p_{\text{click}}(1 - h(E_Z) - h(E_X)). \qquad (6.11)$$

By optimizing the key rate over the input parameter q, one obtains an optimal value in the range: $q \in [0.88, 0.94]$ for every value of η. This increases the weight of the desired contribution $|\psi_{k_d 1}\rangle\langle\psi_{k_d 1}|_{AB}$ in the parties' shared state (6.7), as explained above.

The overall scaling of the key rate (6.11) with respect to η can be immediately visualized by neglecting the terms of second order in $(1-q)$ (which are small when q is optimized). In this approximation we have that $E_X \approx E_Z \approx 0$ and that $2p_{\text{click}} \approx 2q(1-q)\sqrt{\eta}$, hence the key rate scales with $\sqrt{\eta}$, as anticipated.

6.3.2   Actual Protocol

Here we present the actual TF-QKD protocol introduced in [13], which is inspired by the idealized protocol above but it is much more practical to implement. First of all, note that the measurements performed by Alice and Bob commute with the operations of the relay. This means that the measurements in step 5 can be performed right after step 1, i.e. the parties can directly measure their qubit after generating the entangled state (6.4). In doing so, we turn the protocol into a prepare-and-measure scheme where Alice, upon choosing the X basis, prepares an optical pulse a in the state:

$$|X_{b_A}\rangle_a := \sqrt{q}|0\rangle_a + (-1)^{b_A}\sqrt{1-q}|1\rangle_a, \qquad (6.12)$$

depending on the value of a bit b_A, chosen at random. Instead, when Alice chooses the Z basis, she prepares the pulse in the Fock state:

$$|Z_{b_A}\rangle_a := |b_A\rangle_a \qquad (6.13)$$

where the vacuum $|0\rangle_a$ ($b_A = 0$) is selected with probability q and the single-photon state $|1\rangle_a$ ($b_A = 1$) is selected with probability $1 - q$. Bob prepares his optical signal in analogous states. The other steps of the protocol remain unchanged.

We remark that this prepare-and-measure scheme is equivalent to the entanglement-based idealized protocol from the point of view of the security and achieved key rate. However, it does not require the generation of entanglement between a local qubit and an optical signal, which might be experimentally demanding. We now replace the states prepared in the current prepare-and-measure scheme with more practical ones, while leaving all the other protocol steps unchanged. In this way we come to the final TF-QKD protocol of [13], which is summed up in Fig. 6.1.

The form of the states (6.12) prepared when the X basis is chosen, combined with the fact that the optimal value for q is close to one, suggest much more practical states to prepare an optical pulse in, namely coherent states $|(-1)^{b_A}\alpha\rangle$ of low intensity $|\alpha|^2$. Indeed, by recalling that a coherent state can be expressed as a superposition of Fock states (5.1), one notices that the states in (6.12) can be well approximated by the WCP:

$$X \text{ basis:} \quad |(-1)^{b_A}\alpha_A\rangle, \qquad (6.14)$$

with an appropriate amplitude α_A and where b_A is a random bit. Bob prepares a coherent state analogous to (6.14) whose amplitude α_B can differ from Alice's.

The Z-basis states (6.13) are Fock states of fixed photon number. Thus, the corresponding error rate E_Z is linked to the probabilities that Alice and Bob send a certain number of photons to the relay and the detection is successful. We have seen (c.f. Sect. 5.2) that such probabilities can be estimated by using the decoy-state method. Now, since the Z-basis rounds do not contribute to key generation, the states prepared in these rounds have the only purpose of quantifying E_Z. Therefore, we can replace them with the more practical phase-randomized WCPs, and estimate E_Z with the decoy-state method. Thus, upon choosing the Z basis, Alice prepares a phase-randomized WCP:

$$Z \text{ basis:} \quad \rho_{\mu_i} = \sum_{n=0}^{\infty} e^{-\mu_i} \frac{\mu_i^n}{n!} |n\rangle\langle n|, \qquad (6.15)$$

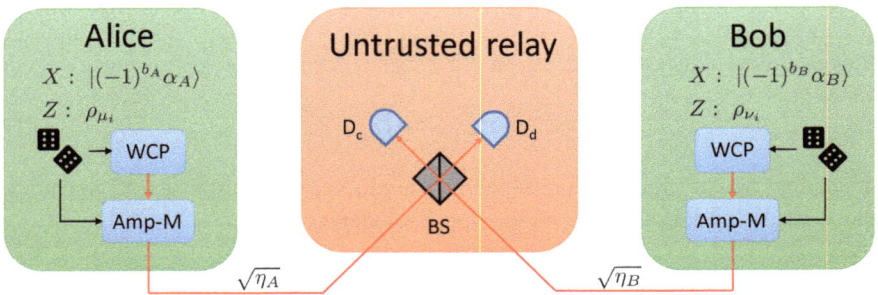

Fig. 6.1 Schematic setup of the practical TF-QKD protocol introduced in [13]. In every round, each party selects the X (Z) basis with probability p_X ($1 - p_X$). When selecting the X basis, Alice (Bob) prepares a WCP whose phase encodes her (his) random key bit b_A (b_B). When the Z basis is selected, she (he) prepares a phase-randomized WCP to implement the decoy state method. Both parties send their pulses to the central relay through channels of transmittance $\sqrt{\eta_A}$ for Alice and $\sqrt{\eta_B}$ for Bob. Here, the incoming pulses are combined into a 50:50 BS followed by two threshold detectors at its output ports. The relay announces the results of the detection k_c, k_d. The parties only keep those rounds in which they chose the same basis and $k_c \oplus k_d = 1$, all the other rounds are discarded. The bits b_A and $b_B \oplus k_d$ form the parties' raw keys

whose intensity μ_i is randomly drawn from a set $\{\mu_i\}$. Analogously, Bob prepares a phase-randomized WCP ρ_{ν_i} with intensity ν_i randomly drawn from the set $\{\nu_i\}$. The two sets of intensities can be different for Alice and Bob.

Remark 6.1 *We stress the fact that the TF-QKD protocol of [13], instead of requiring a global phase post-selection like the original TF-QKD scheme [12], requires a global phase pre-selection which fixes the phases of the coherent states in the X-basis rounds. This can be achieved if the parties share a phase-reference that can also be controlled by Eve. The feasibility of this solution has been experimentally proved [18–20, 22]. Conversely, in the Z-basis rounds the phase-reference is not needed as the parties prepare locally phase-randomized WCPs. This makes the TF-QKD protocol of [13] quite robust against potential phase misalignments, since they only affect the X-basis rounds.*

6.3.3 Error Rates Estimation

When performing the practical TF-QKD protocol outlined above, the quantities that Alice and Bob observe, after revealing their inputs in a fraction of the rounds, are the gains $p_{XX}(k_c, k_d|b_A, b_B)$ and $p_{ZZ}(k_c, k_d|\mu_i, \nu_j)$. The former is the probability that the relay announces the detection pattern k_c, k_d given that Alice (Bob) prepared $|(-1)^{b_A}\alpha_A\rangle$ ($|(-1)^{b_B}\alpha_B\rangle$), while the latter is the probability that the relay announces the detection pattern k_c, k_d given that Alice (Bob) prepared ρ_{μ_i} (ρ_{ν_j}).

From the observed gains, the parties can estimate the error rates E_X and E_Z as follows. In the following, we assume that the detection pattern k_c, k_d is such that $k_c \oplus k_d = 1$.

E_X **estimation** From Bayes' theorem [23] we obtain:

$$p_{XX}(b_A, b_B | k_c, k_d) = \frac{1}{4} \frac{p_{XX}(k_c, k_d | b_A, b_B)}{p_{XX}(k_c, k_d)}, \tag{6.16}$$

where:

$$p_{XX}(k_c, k_d) = \frac{1}{4} \sum_{b_A, b_B = 0}^{1} p_{XX}(k_c, k_d | b_A, b_B). \tag{6.17}$$

We can then compute the error rate E_X in (6.5), for the detection pattern k_c, k_d, as follows:

$$E_X^{k_c, k_d} = \sum_{j=0}^{1} p_{XX}(b_A = j, b_B = j \oplus k_c | k_c, k_d), \tag{6.18}$$

where the probabilities $p_{XX}(b_A, b_B | k_c, k_d)$ are given in (6.16).

E_Z **estimation** What we want is an estimation of the error rate E_Z that characterizes the rounds where Alice and Bob chose the X basis. Suppose that, upon choosing the X basis, Alice (Bob) implements and entanglement-based version of the TF-QKD protocol. Namely, she (he) prepares the following entangled state $|\Phi\rangle_{Aa}$ ($|\Phi\rangle_{Bb}$) between the qubit A (B) and the WCP:

$$|\Phi\rangle_{Aa} = \frac{|+\rangle_A |\alpha_A\rangle_a + |-\rangle_A |-\alpha_A\rangle_a}{\sqrt{2}}, \tag{6.19}$$

and she (he) delays the X measurement on the qubit until the detection at the relay has occurred. Note that Eve cannot distinguish this scenario from the actual scenario in (6.14). The global state of the parties' qubits and signals, after the relay announced outcome k_c, k_d, reads:

$$|\chi^{k_c, k_d}\rangle_{Aa'Bb'} := \frac{M_{a,b}^{k_c, k_d} |\Phi\rangle_{Aa} |\Phi\rangle_{Bb}}{\sqrt{p_{XX}(k_c, k_d)}}, \tag{6.20}$$

where $M_{a,b}^{k_c, k_d}$ is the Kraus operator describing the action of the relay on the signals of Alice and Bob, corresponding to outcome k_c, k_d. The Z-basis error, as defined in (6.6), affecting the X-basis rounds is thus given by:

$$E_Z^{k_c, k_d} = \sum_{j=0}^{1} \left\| {}_{AB}\langle jj | \chi^{k_c, k_d}\rangle_{Aa'Bb'} \right\|^2. \tag{6.21}$$

Now, one can derive an upper bound on (6.21) in terms of the yields $Y_{nm}^{k_c,k_d}$ in the Z basis, i.e. the probability that the relay announces k_c, k_d given that Alice and Bob sent n and m photons, respectively, after choosing the Z basis. The upper bound on E_Z reads [13]:

$$
\bar{E}_Z^{k_c,k_d} := \frac{1}{p_{XX}(k_c,k_d)} \left[\left(\sum_{n,m=0}^{\infty} c_{2n}^A c_{2m}^B \sqrt{Y_{2n\,2m}^{k_c,k_d}} \right)^2 \right.
$$

$$
\left. + \left(\sum_{n,m=0}^{\infty} c_{2n+1}^A c_{2m+1}^B \sqrt{Y_{2n+1\,2m+1}^{k_c,k_d}} \right)^2 \right], \tag{6.22}
$$

where $c_n^{A(B)}$ is defined as: $c_n^{A(B)} = e^{-\frac{\alpha_{A(B)}^2}{2}} \alpha_{A(B)}^n / \sqrt{n!}$. Then, the yields appearing in the bound (6.22) can be estimated with the decoy-state method, by relying on the gains observed in the Z basis: $p_{ZZ}(k_c, k_d|\mu_i, \nu_j)$. Specifically, the yields are constrained by the following set of equations, each corresponding to a particular pair of decoy intensities (μ_i, ν_j):

$$
p_{ZZ}(k_c, k_d|\mu_i, \nu_j) = \sum_{n,m=0}^{\infty} e^{-\mu_i - \nu_j} \frac{\mu_i^n \nu_j^m}{n!m!} Y_{nm}^{k_c,k_d} \quad \mu_i \in \{\mu_i\},\ \nu_j \in \{\nu_j\}, \tag{6.23}
$$

similarly to what we have seen for the decoy-state method applied to the BB84 protocol (5.9). Note that in this case, differently from the usual decoy-state method, one needs to derive *upper* bounds on the yields in (6.22), which correspond to a lower bound on the key rate. Furthermore, since one can only bound a subset of the infinite amount of yields appearing in (6.22), the remaining yields are trivially upper bounded by one.

6.3.4 Secret Key Rate

The asymptotic secret key rate of the practical TF-QKD protocol introduced in [13] is given by:

$$
r_{\mathrm{TF}} \geq r_{\mathrm{TF}}^{1,0} + r_{\mathrm{TF}}^{0,1}, \tag{6.24}
$$

where $r_{\mathrm{TF}}^{k_c,k_d}$ is the contribution due to the detection event k_c, k_d with $k_c \oplus k_d = 1$, defined as:

$$
r_{\mathrm{TF}}^{k_c,k_d} = p_{XX}(k_c, k_d) \left[1 - h(E_X^{k_c,k_d}) - h(\bar{E}_Z^{k_c,k_d}) \right] \tag{6.25}
$$

with $p_{XX}(k_c, k_d)$, $E_X^{k_c,k_d}$ and $\bar{E}_Z^{k_c,k_d}$ given in (6.17), (6.18) and (6.22), respectively. In Fig. 6.2 we plot the asymptotic key rate (6.24) of the practical TF-QKD protocol in [13] as a function of the total distance L between Alice and Bob, assuming that they

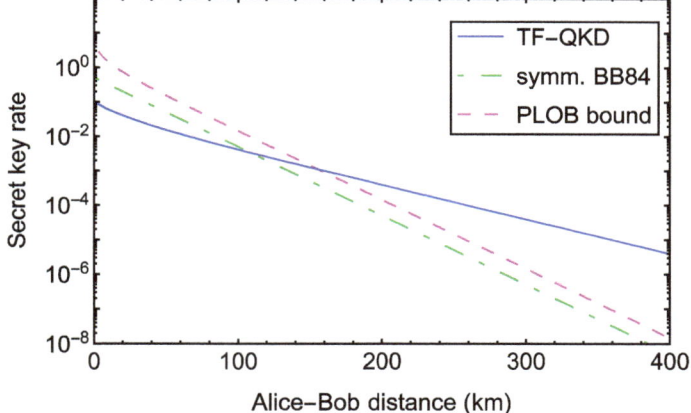

Fig. 6.2 Logarithmic plot of the asymptotic secret key rate of the TF-QKD scheme in [13] (Eq. 6.24, solid blue) and of the symmetric BB84 protocol with a single-photon source (Eq. 6.26, dot-dashed green), as a function of the distance between Alice and Bob. We assume that the quantum channels are telecom fibres with $0.2\,\text{dB km}^{-1}$ of loss. Apart from this, the implementation of both protocols is error-free. We also plot the PLOB bound (Eq. 6.1, dashed magenta). We observe the square-root improvement in the scaling of the TF key rate with the transmittance, compared to the BB84 protocol and to the PLOB bound. Note that a square-root improvement results in an increased slope in this logarithmic plot

are both linked to the central relay by equally-long telecom fibres with $0.2\,\text{dB km}^{-1}$ of loss. Hence the transmittances of their channels are: $\sqrt{\eta_A} = \sqrt{\eta_B} =: \sqrt{\eta} = 10^{-\frac{0.2L}{20}}$. In Fig. 6.2 we also report the PLOB bound (6.1) and the asymptotic key rate of a symmetric BB84 protocol (c.f. Sect. 3.2) implemented with a single-photon source:

$$r_{\text{symBB84}} = \eta/2. \qquad (6.26)$$

For both protocols, we assume an ideal implementation where the only source of errors is the loss in the quantum channel(s) (simulations including other sources of error can be found in [13]). In the case of the TF scheme, we also assumed an infinite number of decoy intensity settings, which basically means that the parties know the exact values of all the yields appearing in the upper bound $\bar{E}_Z^{k_c,k_d}$. Finally, we optimized the TF key rate (6.24) over the *signal intensities* α_A^2 and α_B^2 of the WCPs prepared by Alice and Bob in the X-basis rounds (for simplicity $\alpha_A, \alpha_B \in \mathbb{R}$). Note that, for identical losses $\sqrt{\eta_A} = \sqrt{\eta_B}$, the optimal signal intensities are equal: $\alpha_A^2 = \alpha_B^2$.

From Fig. 6.2 we observe the improved scaling of the TF key rate compared to the BB84 protocol and to the upper bound on the key rate of any point-to-point QKD protocol, i.e. the PLOB bound. Indeed, while the TF key rate scales with $\sim \sqrt{\eta}$, the BB84 protocol and the PLOB bound scale with $\sim \eta$. In particular, there exists a loss threshold/distance after which TF-QKD performs better than any point-to-point QKD scheme, i.e. when the PLOB bound is surpassed. We emphasize that this theoretical prediction has been recently confirmed experimentally [18–22].

6.4 Twin-Field QKD with Finite Decoys and Asymmetric Channels

In the paper that introduced the TF-QKD protocol without phase post-selection [13], the key rate performance is mainly investigated in the unrealistic scenario where Alice and Bob can use an infinite number of decoy intensity settings.

In order to investigate the real performance of the proposed TF scheme, in [24] we derive analytical bounds on several yields appearing in the upper bound (6.22) on the error rate E_Z, when the parties have at their disposal either two, three, or four decoy intensity settings each. These are the most relevant cases from an experimental point of view. Since to every pair of decoy intensities corresponds a linear constraint on the yields (see (6.23)), increasing the number of decoy intensities enables us to bound a larger number of yields and more tightly. In the limit of infinitely many decoy intensity settings, the parties can correctly estimate all the infinite yields appearing in (6.22). Moreover, the larger the number of yields with an analytical bound, the smaller the number of yields trivially bounded by one in (6.22). This has the obvious effect of increasing the protocol's key rate.

Furthermore, the yields bounds enable a closed analytical expression of the secret key rate, which is particularly useful when optimizing the protocol's performance over a large set of parameters, e.g., in the finite-key regime.

Equipped with the derived bounds on the yields, we show that two decoy settings are enough to beat the PLOB bound (6.1) and that four decoy settings are close to optimal, i.e. the resulting key rate is almost indistinguishable from that where Alice and Bob have infinite decoy settings.

In the performance analysis conducted in [13] it is also assumed that the losses affecting the quantum channels of Alice and Bob are equal and so are the optimal signal and decoy intensities. However, this does not reflect realistic scenarios where two parties establishing a secret key, e.g., in the context of a quantum network, are likely to be at different distances from the untrusted relay which processes their signals according to the TF-QKD protocol in [13]. Moreover, potential intensity fluctuations affecting the parties' lasers are likely to be uncorrelated, causing the parties to effectively employ different signal and decoy intensities.

In order to address these issues, in [25] we investigate the performance of the TF-QKD protocol of [13] in asymmetric-loss scenarios and in the presence of independent laser intensity fluctuations. To this aim, we derive new analytical bounds on the relevant yields appearing in (6.22), when the parties use two independent sets of decoy intensities ($\{\mu_i\}$ and $\{\nu_j\}$). In particular, based on the results of [24], we consider the cases of two, three and four decoy intensity settings for each party.

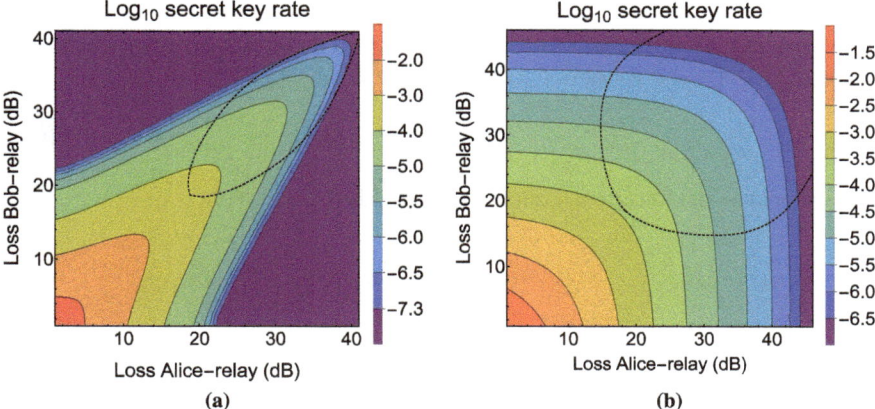

Fig. 6.3 Contour lines for the asymptotic secret key rate of the TF-QKD scheme in [13], evaluated with the yields bounds derived in [25] relative to three decoy intensity settings. The key rate is optimized over the signal α_A^2, α_B^2 and decoy intensities $\{\mu_i\}$, $\{\nu_i\}$ of Alice and Bob, respectively. In **a** the additional constraints are: $\alpha_A = \alpha_B$ and $\{\mu_i\} = \{\nu_i\}$. We observe that, when the parties can use asymmetric intensities **b**, the key rate is never enhanced by adding noise in one of the channels in order to symmetrize the losses. The black dotted lines enclose the region where the key rate beats the PLOB bound (6.1). The utilized channel model comprises a 2% polarization misalignment, a 2% phase misalignment and a dark count probability in each detector of 10^{-7}

We then numerically optimize the key rate over the (potentially) different signal and decoy intensities.

An example of the advantage gained by allowing Alice and Bob to independently select their signal and decoy intensities is given in Fig. 6.3. Here, we provide two contour plots of the secret key rate optimized over the signal and decoy intensities, as a function of the loss (measured in dB) in the quantum channels linking Alice and Bob to the untrusted relay. For instance, if Loss$_A$ is the loss in Alice's channel, then the transmittance of her channel is given by: $\sqrt{\eta_A} = 10^{-\text{Loss}_A/10}$.

The plot in Fig. 6.3a is optimized with the constraint that Alice and Bob use the same set of decoy intensities and the same signal intensity, while the plot in Fig. 6.3b is optimized without that constraint—i.e. Alice and Bob are free to independently optimize their intensities. We observe a drastic improvement of the key rate when the parties can independently select the signal and decoy intensities, especially when the losses in two channels are highly asymmetric. Surprisingly, when the parties are forced to employ the same intensities and their losses are significantly asymmetric, it is convenient for them to artificially increase the loss in one of their channels (e.g., by adding fibre) in order to maximize the key rate (see Fig. 6.3a).

In [25] we also show that the TF-QKD protocol of [13] is considerably robust against independent intensity fluctuations of the parties' lasers.

Finally we illustrate, in the simplest case of two decoy intensities per party, the procedure we adopt in [24, 25] to derive good bounds on the relevant yields in (6.22). In particular, as an example we derive the upper bound on $Y_{11}^{k_c,k_d}$. We assume that

Alice (Bob) can choose among the decoy intensities $\{\mu_0, \mu_1\}$ with $\mu_0 > \mu_1$ ($\{v_0, v_1\}$ with $v_0 > v_1$). To keep the notation simple, we define the following rescaled gains in the Z basis (where we omit the detection pattern k_c, k_d):

$$\tilde{Q}^{\mu_i, v_j} = e^{\mu_i + v_j} p_{ZZ}(k_c, k_d | \mu_i, v_j), \tag{6.27}$$

and we rewrite the linear constraints on the yields (6.23) as follows:

$$\tilde{Q}^{\mu_i, v_j} = \sum_{n,m=0}^{\infty} \frac{Y_{nm}}{n!m!} \mu_i^n v_j^m. \tag{6.28}$$

Consider the following combination of gains:

$$G := \tilde{Q}^{0,0} + \tilde{Q}^{1,1} - \tilde{Q}^{0,1} - \tilde{Q}^{1,0} = \sum_{n,m=0}^{\infty} \frac{Y_{nm}}{n!m!} (\mu_0^n - \mu_1^n)(v_0^m - v_1^m), \tag{6.29}$$

and note that the coefficients of the yields Y_{n0} and Y_{0m} are null for every n and m. We can then recast the last expression as follows:

$$G = Y_{11}(\mu_0 - \mu_1)(v_0 - v_1) + \sum_{\substack{n,m=1 \text{ s.t.} \\ n+m>2}}^{\infty} \frac{Y_{nm}}{n!m!} \left(\mu_0^n - \mu_1^n\right) \left(v_0^m - v_1^m\right). \tag{6.30}$$

We emphasize that Y_{11} is now the yield with the largest coefficient[2] in (6.30), thus trivially bounding the other yields is not as harmful as it would be if they had the largest coefficients.

An upper bound on Y_{11} is then obtained by considering the worst-case scenario for the other yields, taking into account that they are probabilities, i.e. $0 \le Y_{nm} \le 1$. Since all the yields' coefficients have the same sign in (6.30), the yield Y_{11} is maximal when all the other yields are minimal. Hence, the upper bound on Y_{11} is obtained by setting all the other yields to zero in (6.30):

$$Y_{11}^U = \min \left\{ \frac{G}{(\mu_0 - \mu_1)(v_0 - v_1)}, 1 \right\}. \tag{6.31}$$

Note that by taking the minimum in the above expression we make sure that Y_{11}^U is a meaningful bound on a probability. The upper bound in (6.31) is only expressed in terms of input parameters (the decoy intensities) and observed gains contained in G (see (6.29)).

---

[2]Optimal decoy intensity values are typically smaller than one, and one of the decoy intensities of each party is always as small as allowed by the experimental equipment [24, 25].

6.5 Conference Key Agreement with Single Photon Interference

The founding idea of TF-QKD, elucidated in Sect. 6.3.1, can also be generalized with some adjustments to the multiparty scenario. Indeed, in [26] we devise a conference key agreement (CKA) where N parties simultaneously distil a secret conference key through single-photon interference occurring at an untrusted relay.

In particular, parties Alice$_1$, Alice$_2$, ..., Alice$_N$ establish the conference key by sending optical pulses to the relay and by performing suitable measurements on their qubits. The resulting CKA is sketched in Fig. 6.4 and each round is characterized by the following steps.

1. Alice$_i$ ($i = 1, \ldots, N$) prepares an optical pulse a_i entangled with a qubit A_i she holds:

$$|\Phi\rangle_{A_i a_i} = \sqrt{q}|0\rangle_{A_i}|0\rangle_{a_i} + \sqrt{1-q}|1\rangle_{A_i}|1\rangle_{a_i} \quad 0 \leq q \leq 1 \qquad (6.32)$$

where $|0\rangle_{a_i}, |1\rangle_{a_i}$ are the photon's vacuum and single-photon state, while $\{|0\rangle_{A_i}, |1\rangle_{A_i}\}$ is the computational basis of qubit A_i (Z basis).

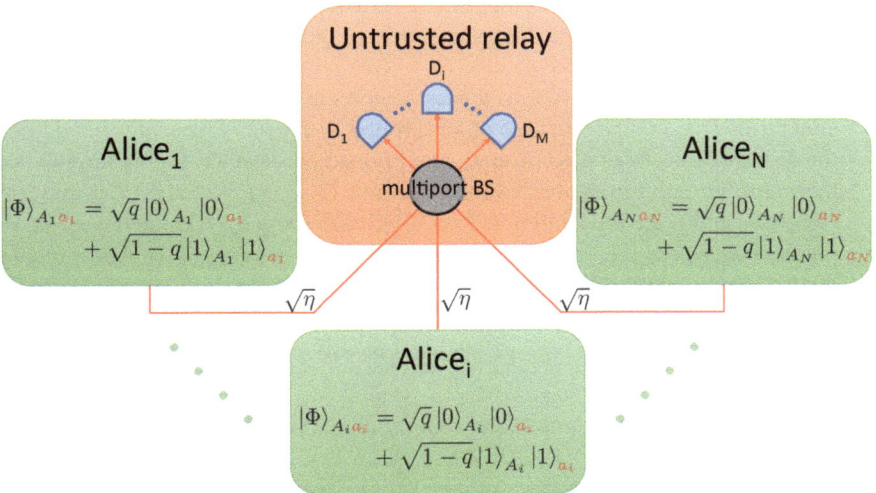

Fig. 6.4 N-party CKA based on single-photon interference [26]. Every party initially prepares an entangled state $|\Phi\rangle_{A_i a_i}$ (6.32) between a qubit she holds and an optical signal. The state is unbalanced towards the vacuum: $1 - q \ll 1$. The signals are then sent to the untrusted relay through optical channels with transmittance $\sqrt{\eta}$. The relay combines the pulses in a multiport BS with M inputs and M outputs ($M \geq N$) featuring a detector at every output port, and then announces the outcome of the detection of each detector. The events in which only one detector clicked are most likely caused by the detection of just one photon, sent by one of the parties with equal probability. Hence, the conditional state of the qubits $A_1, \ldots, A_N$ is well approximated by a W state, which can be used by the parties to distil a conference key

2. Every party sends her optical pulse a_i to the relay via optical channels of transmittance $\sqrt{\eta}$. The transmittance between any two parties is thus η.
3. The relay applies a Bell-multiport beam splitter [27–30] with M input and M output ports (where $M \geq N$) to the incoming pulses followed by a threshold detector D_i ($i = 1, \ldots, M$) at each output port. If $M > N$, some inputs ports receive the vacuum. The action of the multiport beam splitter (BS) is defined by the following unitary transformation which reduces to the standard 50:50 BS for $M = 2$:

$$a_{\text{in},i}^\dagger \mapsto \sum_{j=1}^{M} U_{ij} a_{\text{out},j}^\dagger, \tag{6.33}$$

where $a_{\text{in},i}^\dagger$ ($a_{\text{out},j}^\dagger$) are the creation operators of the incoming (outgoing) photons and U_{ij} are the coefficients of a unitary matrix defined as:

$$U_{ij} = \frac{1}{\sqrt{M}} e^{i\frac{2\pi}{M}(i-1)(j-1)} \qquad i, j \in \{1, \ldots, M\}. \tag{6.34}$$

4. The relay broadcasts the outcome k_j of every detector D_j, with $k_j = 0$ ($k_j = 1$) corresponding to a no-click (click) event. The round is discarded when $\sum_{j=1}^{M} k_j \neq 1$, i.e. whenever single-photon interference did not occur. The probability that only detector D_j clicked is p_D and is independent of the detector due to the symmetric action of the multiport BS.
5. The round is classified either as a parameter-estimation (PE) round or as a key-generation (KG) round.[3] In case of a PE round, every party measures her qubit in the Z-basis. In case of a KG round, conditioned on detector D_j clicking, Alice$_i$ measures her qubit in the basis of the operator $O_{XY}(\varphi_i) = \cos(\varphi_i)X + \sin(\varphi_i)Y$ (where X and Y are the Pauli operators), with $\varphi_i = \arg(U_{ij})$. Upon observing the outcome x (where $x = \pm 1$ are the eigenvalues of Z and $O_{XY}(\varphi_i)$), Alice$_i$ records the bit value b_i with $(-1)^{b_i} = x$.
6. The bits b_i measured in KG rounds form Alice$_i$'s raw key, while those obtained in PE are used to detect Eve's action and quantify the information she gained on Alice$_1$'s raw key.

After performing a suitable number of rounds, all the parties perform one-way error correction to match their raw keys to Alice$_1$'s raw key. The parties then perform privacy amplification on their error-corrected keys to distil a shorter, secret, conference key.

The error rates E_Z and $E_{A_1 A_i}$ devoted to quantify Eve's knowledge and the information that Alice$_1$ needs to send for error correction are defined as follows:

---

[3]The parties can know in advance the classification of each round from a preshared key they hold, see Remark 4.2.

$$E_Z = p_{PE}\left[b_1 = \bigoplus_{i=2}^{N} b_i \,\middle|\, \sum_{j=1}^{M} k_j = 1\right] \tag{6.35}$$

$$E_{A_1 A_i} = p_{KG}\left[b_1 \neq b_i \,\middle|\, \sum_{j=1}^{M} k_j = 1\right], \tag{6.36}$$

where $p_{PE}[\Omega]$ ($p_{KG}[\Omega]$) is the probability that the event Ω occurred in a PE (KG) round. The asymptotic secret conference key rate achieved by the described CKA protocol based on single-photon interference events reads:

$$r_{spCKA} = M p_D \left(1 - h(E_Z) - \max_{2 \le i \le N} h(E_{A_1 A_i})\right) \tag{6.37}$$

Notably, the above CKA scheme and the relative asymptotic key rate reduce to the idealized TF-QKD protocol presented in Sect. 6.3.1 when $N = M = 2$.

6.5.1   Multipartite QKD with a W State

Here we would like to provide, as in the case of the idealized TF-QKD protocol, a bit of intuition on why the above CKA protocol works.

When optimizing the conference key rate in (6.37) over q, we obtain values such that $1 - q \ll 1$, that is the initial entangled states (6.32) are strongly unbalanced towards the vacuum. Therefore, the rounds where only one detector clicked are mainly caused by the sending and detection of just one photon. The photon could be sent by any Alice$_i$, implying that her qubit would be in state $|1\rangle_{A_i}$ while the qubits of the other parties would be in state $|0\rangle_{A_{\neq i}}$. Since the multiport BS creates a coherent superposition of all these possibilities, by post-selecting the rounds where e.g., detector D_j clicked, the state of the parties' qubits is approximately given by the following W-class state[4] [31]:

$$|W_j\rangle_{A_1 \dots A_N} = \frac{1}{\sqrt{N}} \sum_{i=1}^{N} \sqrt{M} U_{ij} |\mathbf{v}_i\rangle, \tag{6.38}$$

where the bitstring $\mathbf{v}_i \in \{0, 1\}^N$ is composed of all zeroes except for the i-th bit which has value one and where U_{ij} is defined in (6.34). We remark that the qubits' state in (6.38) is only an approximation in the limit $1 - q \ll 1$ where terms of second or higher order in $1 - q$ have been neglected and in the ideal scenario of no losses ($\sqrt{\eta} = 1$) and no other sources of noise.

Although it is proven [32] that no N-qubit state other than the GHZ state (4.1) can lead to perfectly correlated outcomes in one measurement basis (for $N \ge 3$), the parties can still distil a secret conference key by properly measuring their qubits in state (6.38).

---

[4]The W state usually considered in the literature reads: $\frac{1}{\sqrt{N}} \sum_{i=1}^{N} |\mathbf{v}_i\rangle$.

Indeed, the measurements that the parties perform in KG rounds (see protocol description) are chosen to minimize the key-bit error rate $E_{A_1 A_i}$ between Alice$_1$ and any other party. In particular, one could view the measurement of Alice$_i$ in the eigenbasis of $\cos[\arg(U_{ij})]X + \sin[\arg(U_{ij})]Y$ as composed of two steps. First she rotates her X operator in the (x, y)-plane of the Bloch sphere by an angle $\arg(U_{ij})$, in order to remove the effect of the complex phase $\sqrt{M}U_{ij}$ introduced by the BS when D_j clicked. Then she measures in the eigenbasis of the rotated operator.

The resulting error rate $E_{A_1 A_i}$ (6.36) the parties would observe if their qubits were exactly in the state (6.38) is given by:

$$E_{A_1 A_i} = \frac{1}{2} - \frac{1}{N}. \tag{6.39}$$

This intrinsic error rate affecting the parties' raw key bits is unavoidable due to the fact that they are measuring a W-class state, instead of a GHZ state. Conversely, the error rate E_Z (6.35) computed in PE is null on the state (6.38), confirming that in ideal conditions Eve does not gain any information.

We have thus argued that multipartite QKD can also be implemented on a W state, instead of the conventional GHZ state used in the majority of cases, e.g., with the multiparty BB84 and six-state protocols (c.f. Chap. 4). Despite presenting the drawback of the intrinsic error rate (6.39), the CKA based on the W state becomes dramatically advantageous in high-loss scenarios.

Indeed, the W state is post-selected when single-photon interference occurred at the relay. This implies that the resulting conference key rate (6.37) scales linearly with the transmittance $\sqrt{\eta}$ of one of the channels linking the parties to the relay.

Let us now consider a generic optical implementation of a CKA based on an N-qubit GHZ state, where the qubit state is encoded in one of the photon's degrees of freedom (e.g., the polarization). The N photons described by an N-qubit GHZ state are distributed from a central untrusted node to the N parties through the same quantum channels with transmittance $\sqrt{\eta}$. The conference key rate of this protocol cannot scale better than $\sim (\sqrt{\eta})^N$, since all the photons are required to arrive in order to have a successful round.

Clearly, in a high-loss scenario ($\sqrt{\eta} \to 0$) the conference key rate of our CKA based on single-photon interference will outperform any CKA based on GHZ states implemented as described above.

Remark 6.2 (*Impossibility of prepare-and-measure CKA*). *We emphasize that the measurements performed by the parties in the KG rounds do not commute with the operations of the relay, inasmuch as they depend on which detector clicked. This means that the CKA cannot be turned into a prepare-and-measure scheme where each party prepares some optical signal depending on a random bit and on the basis choice. For this reason, it cannot be regarded as an MDI-QKD protocol since the parties still need to perform trusted measurements on their qubits.*

This contrasts with the TF-QKD idealized protocol presented in Sect. 6.3.1, which is recovered here for $N = M = 2$. The bipartite case is special since the complex *phase introduced by the BS in the shared state (6.38) reduces to a minus sign, which can be reabsorbed by asking Alice$_2$ to flip her classical outcome b_2 when detector D_2 clicks, as described in Sect. 6.3.1. This removes the need to adjust the parties' measurements depending on the result of the detection, hence making these two steps commute.*

Nevertheless, the quantum operations of the CKA scheme in [26] seem to be feasible with present-day technology. In particular, the parties' qubits may be realized with electronic spin-1 systems of nitrogen-vacancy (NV) centres in diamond [33, 34], characterized by second-long coherence times [35]. This would allow the parties to delay the measurements on their qubits until the relay announces which detectors clicked, as requested by the CKA protocol. Moreover, the qubit state can be accurately tuned by shining microwave pulses and can be subsequently entangled to the photon number (presence or absence of a photon) by exciting its ground state, which then spontaneously emits a photon [33]. This process realizes the qubit-photon entangled state (6.32).

6.5.2 Performance Assessment

In [26], we prove the CKA security in the finite-key scenario for the most general attacks the eavesdropper can perform. We also investigate the protocol's performance for a realistic channel model that accounts for polarization and phase misalignments and dark counts in the detectors.

In order to benchmark the performance of our CKA based on a central untrusted relay, we consider a scenario where the relay is removed and the parties are all connected in a star network where the transmittance between any two parties is η. In this configuration, we consider the conference key rate generated by the following strategy and compare it with the CKA key rate (6.37). One special party, say Alice$_1$, performs the best possible bipartite QKD protocol with every other party, thus establishing $N-1$ secret keys whose key rate is given by the PLOB bound (6.1). Alice$_1$ then uses the established keys to distribute the conference key to the other parties with one-time pad encryption. The resulting conference key rate is thus given by the rate at which the bipartite keys were generated, rescaled by the factor $1/(N-1)$ which accounts for the fact that Alice$_1$ repeated the bipartite scheme $N-1$ times. The conference key rate resulting from the above strategy implemented on the star network reads:

$$r_{\text{dir.tr.}} = \frac{-\log_2(1-\eta)}{N-1} \tag{6.40}$$

and we call it the *direct-transmission* bound. This is similar to what is done for TF-QKD when benchmarked against the PLOB bound (c.f. Fig. 6.2), which bounds the highest possible key rate achieved between Alice and Bob if the untrusted relay is removed.

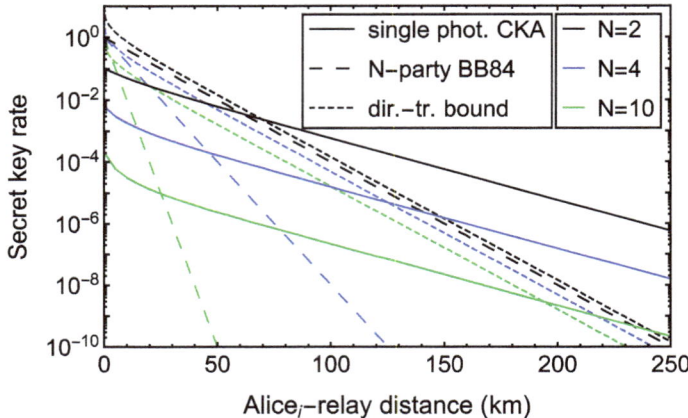

Fig. 6.5 Conference key rates yielded by the CKA based on single-photon interference (solid lines, Eq. 6.37 optimized over q with $M = N$) and by the N-party BB84 protocol (dashed lines, Eq. 6.41) as a function of the channel length between one party and the untrusted relay, for different numbers of parties N (black, blue, green). The dotted lines are the direct-transmission bound (Eq. 6.40). The experimental setup is assumed to be ideal except for the lossy quantum channels with 0.2 dB km^{-1} of loss. The improved key rate scaling of the single-photon-based CKA enables it to outperform both the N-BB84 protocol and the direct-transmission bound on longer distances

In Fig. 6.5 we plot the CKA key rate in (6.37) (solid lines) and the conference key rate of the N-partite BB84 protocol (dashed lines), as a function of the distance between one party and the relay and for different numbers of parties ($N = 2, 4, 10$). In Fig. 6.5 we also plot the direct-transmission bound derived in (6.40) (dotted lines).

The conference key rates are obtained in an ideal experimental setup where the only source of errors is the photon loss in the quantum channels. We assumed as usual 0.2 dB km^{-1} of loss in each quantum channel, the typical loss of standard telecom fibre. In [26] we account for more realistic channel models, which include dark counts in the detectors and misalignments of the phase and polarization.

The considered N-BB84 protocol is such that the relay has the function of distributing the entangled photon state to the N parties. In the chosen ideal setting, the conference key rate of the N-BB84 protocol is just given by the probability that each photon reaches the corresponding party:

$$r_{\text{NBB84}} = \eta^{N/2}. \tag{6.41}$$

The CKA key rate (6.37) has been optimized over the parameter q and we fixed the number of BS ports to match the number of parties: $M = N$. We remark that the optimal number of BS ports—and thus detectors—is $M \approx N$ but it actually depends on the loss. Indeed, a larger number of BS ports decreases the possibility of detecting two photons in the same detector, which is a source of error especially at low losses. However, when accounting for dark counts in the detectors, increasing the number of detectors implies a higher probability of dark counts, which is another source of error manifesting itself at high losses with a drop of the key rate.

As anticipated, Fig. 6.5 clearly shows the significant improvement in the conference key rate when employing our CKA based on single-photon interference, as opposed to employing a GHZ-state-based CKA like the N-party BB84 protocol. Moreover, due to the improved scaling with the loss, the CKA key rate (6.37) also surpasses the direct-transmission bound (6.40) for sufficiently long distances, similarly to what happens with the TF-QKD protocol and the PLOB bound (see Fig. 6.2).

References

1. Takeoka, M., Guha, S., & Wilde, M. M. (2014). Fundamental rate-loss tradeoff for optical quantum key distribution. *Nature Communications*, *5*(1), 5235.
2. Pirandola, S., Laurenza, R., Ottaviani, C., & Banchi, L. (2017). Fundamental limits of repeaterless quantum communications. *Nature Communications*, *8*(1), 15043.
3. Briegel, H.-J., Dür, W., Cirac, J. I., & Zoller, P. (1998). Quantum repeaters: The role of imperfect local operations in quantum communication. *Physical Review Letters*, *81*, 5932–5935.
4. Sangouard, N., Simon, C., de Riedmatten, H., & Gisin, N. (2011). Quantum repeaters based on atomic ensembles and linear optics. *Reviews of Modern Physics*, *83*, 33–80.
5. Duan, L.-M., Lukin, M. D., Cirac, J. I., & Zoller, P. (2001). Long-distance quantum communication with atomic ensembles and linear optics. *Nature*, *414*(6862), 413–418.
6. Mazurek, P., Grudka, A., Horodecki, M., Horodecki, P., Łodyga, J., Pankowski, L., et al. (2014). Long-distance quantum communication over noisy networks without long-time quantum memory. *Physical Review A*, *90*, 062311.
7. Munro, W. J., Stephens, A. M., Devitt, S. J., Harrison, K. A., & Nemoto, K. (2012). Quantum communication without the necessity of quantum memories. *Nature Photonics*, *6*(11), 777–781.
8. Azuma, K., Tamaki, K., & Lo, H.-K. (2015). All-photonic quantum repeaters. *Nature Communications*, *6*(1), 6787.
9. Panayi, C., Razavi, M., Ma, X., & Lütkenhaus, N. (2014). Memory-assisted measurement-device-independent quantum key distribution. *New Journal of Physics*, *16*(4), 043005.
10. Abruzzo, S., Kampermann, H., & Bruß, D. (2014). Measurement-device-independent quantum key distribution with quantum memories. *Physical Review A*, *89*, 012301.
11. Azuma, K., Tamaki, K., & Munro, W. J. (2015). All-photonic intercity quantum key distribution. *Nature Communications*, *6*(1), 10171.
12. Lucamarini, M., Yuan, Z. L., Dynes, J. F., & Shields, A. J. (2018). Overcoming the rate-distance limit of quantum key distribution without quantum repeaters. *Nature*, *557*(7705), 400–403.
13. Curty, M., Azuma, K., & Lo, H. -K. (2019). Simple security proof of twin-field type quantum key distribution protocol. *npj Quantum Information*, *5*(1), 64.
14. Cui, C., Yin, Z.-Q., Wang, R., Chen, W., Wang, S., Guo, G.-C., et al. (2019). Twin-field quantum key distribution without phase postselection. *Physical Review Applied*, *11*, 034053.
15. Lin, J., & Lütkenhaus, N. (2018). Simple security analysis of phase-matching measurement-device-independent quantum key distribution. *Physical Review A*, *98*, 042332.
16. Ma, X., Zeng, P., & Zhou, H. (2018). Phase-matching quantum key distribution. *Physical Review X*, *8*, 031043.
17. Wang, X.-B., Yu, Z.-W., & Hu, X.-L. (2018). Twin-field quantum key distribution with large misalignment error. *Physical Review A*, *98*, 062323.
18. Wang, S., He, D.-Y., Yin, Z.-Q., Lu, F.-Y., Cui, C.-H., Chen, W., et al. (2019). Beating the fundamental rate-distance limit in a proof-of-principle quantum key distribution system. *Physical Review X*, *9*, 021046.

19. Liu, Y., Yu, Z.-W., Zhang, W., Guan, J.-Y., Chen, J.-P., Zhang, C., et al. (2019). Experimental twin-field quantum key distribution through sending or not sending. *Physical Review Letters*, *123*, 100505.
20. Chen, J.-P., Zhang, C., Liu, Y., Jiang, C., Zhang, W., Hu, X.-L., et al. (2020). Sending-or-not-sending with independent lasers: Secure twin-field quantum key distribution over 509 km. *Physical Review Letters*, *124*, 070501.
21. Zhong, X., Hu, J., Curty, M., Qian, L., & Lo, H.-K. (2019). Proof-of-principle experimental demonstration of twin-field type quantum key distribution. *Physical Review Letters*, *123*, 100506.
22. Minder, M., Pittaluga, M., Roberts, G. L., Lucamarini, M., Dynes, J. F., Yuan, Z. L., et al. (2019). Experimental quantum key distribution beyond the repeaterless secret key capacity. *Nature Photonics*, *13*(5), 334–338.
23. Stuart, A., & Ord, J. K. (1994). *Kendall's Advanced Theory of Statistics* (Vol. 1). Distribution Theory: Edward Arnold Publishers.
24. Grasselli, F., & Curty, M. (2019). Practical decoy-state method for twin-field quantum key distribution. *New Journal of Physics*, *21*(7), 073001.
25. Grasselli, F., Navarrete, Á., & Curty, M. (2019a). Asymmetric twin-field quantum key distribution. *New Journal of Physics*, *21*(11), 113032.
26. Grasselli, F., Kampermann, H., & Bruß, D. (2019b). Conference key agreement with single-photon interference. *New Journal of Physics*, *21*(12), 123002.
27. Żukowski, M., Zeilinger, A., & Horne, M. A. (1997). Realizable higher-dimensional two-particle entanglements via multiport beam splitters. *Physical Review A*, *55*, 2564–2579.
28. Lim, Y. L., & Beige, A. (2005). Multiphoton entanglement through a bell-multiport beam splitter. *Physical Review A*, *71*, 062311.
29. Peruzzo, A., Laing, A., Politi, A., Rudolph, T., & O'Brien, J. L. (2011). Multimode quantum interference of photons in multiport integrated devices. *Nature Communications*, *2*(1), 224.
30. Spagnolo, N., Vitelli, C., Aparo, L., Mataloni, P., Sciarrino, F., Crespi, A., et al. (2013). Three-photon bosonic coalescence in an integrated tritter. *Nature Communications*, *4*(1), 1606.
31. Dür, W., Vidal, G., & Cirac, J. I. (2000). Three qubits can be entangled in two inequivalent ways. *Physical Review A*, *62*, 062314.
32. Epping, M., Kampermann, H., Macchiavello, C., & Bruß, D. (2017). Multi-partite entanglement can speed up quantum key distribution in networks. *New Journal of Physics*, *19*(9), 093012.
33. Bernien, H., Hensen, B., Pfaff, W., Koolstra, G., Blok, M. S., Robledo, L., et al. (2013). Heralded entanglement between solid-state qubits separated by three metres. *Nature*, *497*(7447), 86–90.
34. Rozpędek, F., Yehia, R., Goodenough, K., Ruf, M., Humphreys, P. C., Hanson, R., et al. (2019). Near-term quantum-repeater experiments with nitrogen-vacancy centers: Overcoming the limitations of direct transmission. *Physical Review A*, *99*, 052330.
35. Abobeih, M. H., Cramer, J., Bakker, M. A., Kalb, N., Markham, M., Twitchen, D. J., et al. (2018). One-second coherence for a single electron spin coupled to a multi-qubit nuclear-spin environment. *Nature Communications*, *9*(1), 2552.

Chapter 7
Device-Independent Quantum Cryptography

Bell's theorem, formulated in 1964, is one of the profound scientific discoveries of the century. Alain Aspect

Abstract The Chapter is organized as follows. In Sect. 7.1 we introduce the concept of non-local correlations and show how they can be witnessed through a Bell inequality violation. We formalize the definitions of different kinds of correlations in Sect. 7.2. In Sect. 7.3 we link the observation of a Bell violation to the security proof of device-independent (DI) protocols. In Sect. 7.4 we describe the bipartite DIQKD protocol based on the Clauser-Horne-Shimony-Holt (CHSH) inequality and prove its security in Sect. 7.5. In Sect. 7.6 we generalize the security proof technique to a whole class of multiparty DI protocols. We conclude by discussing the suitability of full-correlator Bell inequalities for DICKA and present a multipartite Bell inequality specifically built for the task of DICKA (Sect. 7.7). In the Appendix of this Chapter (Sects. 7.8 and 7.9) we provide additional details on the security proof of the CHSH-based DIQKD protocol.

We have already seen in Chaps. 5 and 6 how imperfections in the quantum devices employed in a QKD protocol, when not accounted for in the security proof, can be exploited by an eavesdropper to spoil the protocol's security. In this context, measurement-device-independent (MDI) QKD and twin-field (TF) QKD protocols represent possible solutions as they do not require to trust the measurement devices, which can be completely controlled by the eavesdropper, and yet allow to derive a secret key. However, both MDI-QKD and TF-QKD still require to trust the sources held by the parties.

In the previous Chapters we presented QKD protocols where at least some devices in the experimental apparatus need to be trusted. Of course, we could place our trust in such devices more lightheartedly upon deeply characterizing their functioning. However, the characterization process is often challenging and we might not be

© The Author(s), under exclusive license to Springer Nature Switzerland AG 2021
F. Grasselli, *Quantum Cryptography*, Quantum Science and Technology,
https://doi.org/10.1007/978-3-030-64360-7_7

capable or willing to do it. Indeed, in most cases QKD users are laymen who simply want to purchase a service which guarantees a high level of security, without bothering to verify the claimed security.

Quite astonishingly, secure QKD is still possible even when the whole experimental apparatus is untrusted and potentially under the control of the eavesdropper.[1] Indeed, by exploiting the non-local properties of quantum correlations, device-independent (DI) QKD protocols [1–6] and DI conference key agreement (DICKA) protocols [7–10] deliver the same secret key to a group of two or more parties, respectively, where the security of the key is independent of the actual functioning of the employed devices.

In a similar fashion, in DI randomness generation (DIRG) protocols [11–16], the intrinsic randomness generated by quantum mechanical processes is proven to be private upon the observation of certain non-local correlations. Note that secret true randomness is one of the prerequisites of most quantum cryptographic protocols.

7.1 Bell's Theorem

Bell's theorem [17, 18] states that there exist predictions of quantum theory that cannot be explained by any local theory, i.e. a theory based on the assumption of *locality*. In this Section we clarify our definition of locality and prove Bell's theorem. The proof critically relies on the introduction of a *Bell inequality* [18], that is an inequality involving a linear combination of correlators which is satisfied by every local theory but is violated by quantum mechanics.

In the literature one can find several versions of Bell's theorem's proof, leveraging on different assumptions. Here we mainly follow the proofs presented in [19–22] that make use of the Clauser-Horne-Shimony-Holt (CHSH) inequality [23], arguably the most popular Bell inequality.

Let us consider the following Bell experiment, depicted in Fig. 7.1. Two physical systems, which could have interacted in the past, are now far apart and are individually measured by two parties, Alice and Bob. No information is given on the systems, which are thus treated as black boxes. Each box (system) is equipped with two inputs corresponding to the measurement choices of the parties, and generates a binary output upon selecting an input. Hence, the measurement process consists in Alice (Bob) selecting an input $x \in \{0, 1\}$ ($y \in \{0, 1\}$) on her (his) system and collecting the output $a \in \{-1, 1\}$ ($b \in \{-1, 1\}$). We assume that the measurement processes of Alice and Bob are spacelike separated events.

By repeating the experiment several times, the parties can roughly estimate the probability distribution $p(a, b|x, y)$ governing the occurrence of the outcomes a and b, given the inputs x and y. In general, the outcomes recorded by Alice and Bob

---

[1] Minimal requirements on the devices are still in place, such as the isolation of the trusted parties' labs. Without this requirement, the devices could simply broadcast the established secret key upon completing the protocol.

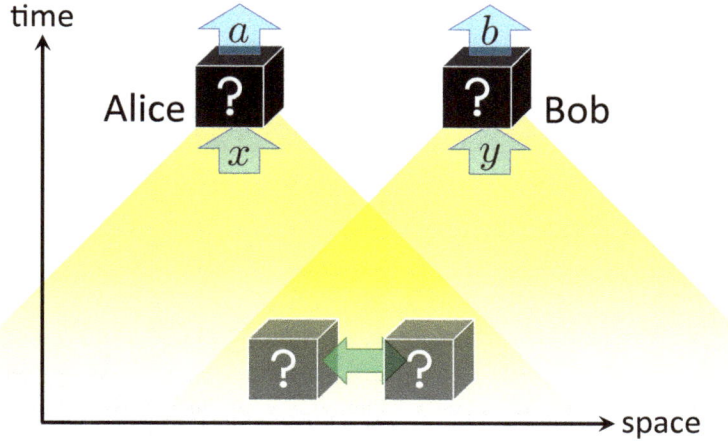

Fig. 7.1 In a bipartite Bell experiment, two unknown systems are given to Alice and Bob. The systems might have interacted in the past, in the spacetime region where their past light cones overlap. Alice (Bob) can only interact with her (his) system by selecting an input x (y) and collecting an output a (b). The interactions of the two parties with the respective systems are assumed to be spacelike separated events

may not be statistically independent, which means that the probability distribution $p(a, b|x, y)$ is not factorized:

$$p(a, b|x, y) \neq p(a|x, y)p(b|x, y). \tag{7.1}$$

This fact could be caused by the previous interaction of the two systems and does not necessarily imply any kind of direct influence of one system on the other. Let us denote with λ the set of underlying (or hidden) variables that completely describe the two systems under consideration. By fixing the value of λ, we fix the *microstate* of the Bell experiment. Hence, λ can account for any dependence relation between the two systems due to their previous interaction. Since the value of λ could vary in different runs of the experiment, the probability distribution $p(a, b|x, y)$ can be expressed as:

$$p(a, b|x, y) = \int d\lambda \, p(a, b|x, y, \lambda)p(\lambda|x, y). \tag{7.2}$$

We remark that so far we did not make any assumption on the theory we employ to describe the Bell experiment, indeed Eq. (7.2) is still completely general. We now assume that any theory describing the experiment should satisfy the following (apparently) natural conditions:

1. (Bell-)locality: All the statistical correlations of the outputs a and b are fully attributable to their past interaction and thus explainable with the knowledge of λ. Formally we have that:

$$p(a, b|x, y, \lambda) = p(a|b, x, y, \lambda)p(b|x, y, \lambda)$$
$$= p(a|x, y, \lambda)p(b|x, y, \lambda), \tag{7.3}$$

where the second equality represents the fact that, conditioned on the knowledge of λ, the residual indeterminacy of a is local and not due to a lack of knowledge of b.

2. <u>Parameter independence</u>: For each microstate λ, the probability of Alice (Bob) obtaining outcome a (b) is independent of the input y (x) selected by Bob (Alice):

$$p(a|x, y, \lambda) = p(a|x, \lambda)$$
$$p(b|x, y, \lambda) = p(b|y, \lambda). \tag{7.4}$$

This assumption is justified by special relativity, according to which spacelike separated measurements do not influence each other's outcome probability distribution.

3. <u>Free will</u>: The measurements inputs are uncorrelated from the underlying state of the systems described by λ:

$$p(x, y, \lambda) = p(x, y)p(\lambda). \tag{7.5}$$

In other words, Alice and Bob are free to choose their inputs independently of the value of the hidden variables λ. By combining this assumption with the previous one, we are basically assuming that spacelike separated parties cannot communicate superluminally, which is the *causality* constraint of relativity.

By employing the assumptions (7.3), (7.4) and (7.5) in (7.2), we obtain the Bell experiment description of a local hidden variable (LHV) model:

$$p(a, b|x, y) = \int d\lambda \, p(\lambda)p(a|x, \lambda)p(b|y, \lambda). \tag{7.6}$$

We will now show that the correlations predicted by quantum mechanics on certain implementations of the Bell experiment cannot be expressed in the form (7.6). To this aim, we define the correlator:

$$\langle a_x b_y \rangle = \sum_{a,b=\pm 1} ab \, p(a, b|x, y), \tag{7.7}$$

as the expectation value of the product of Alice and Bob's outcomes, given that they selected inputs x and y. We then define the CHSH value [23]:

$$S_{\text{CHSH}} = \langle a_0 b_0 \rangle + \langle a_0 b_1 \rangle + \langle a_1 b_0 \rangle - \langle a_1 b_1 \rangle \tag{7.8}$$

and prove that if the probabilities $p(a, b|x, y)$ are explainable in terms of a LHV model (7.6), then the *CHSH inequality* holds:

$$S_{\text{CHSH}} = \langle a_0 b_0 \rangle + \langle a_0 b_1 \rangle + \langle a_1 b_0 \rangle - \langle a_1 b_1 \rangle \leq 2. \tag{7.9}$$

We start by using (7.6) to express the correlators $\langle a_x b_y \rangle$ in (7.7) as a product of local expectation values:

$$\langle a_x b_y \rangle = \int d\lambda \, p(\lambda) \langle a_x \rangle_\lambda \langle b_y \rangle_\lambda \tag{7.10}$$

where $\langle a_x \rangle_\lambda = \sum_a a \, p(a|x, \lambda)$ and similarly for $\langle b_y \rangle_\lambda$, with $\langle a_x \rangle_\lambda, \langle b_y \rangle_\lambda \in [-1, 1]$ (recall that $a, b \in \{-1, 1\}$). By inserting (7.10) into (7.8) we can write that:

$$S_{\text{CHSH}} = \int d\lambda \, p(\lambda) S_{\text{CHSH}}^\lambda, \tag{7.11}$$

where:

$$\begin{aligned} S_{\text{CHSH}}^\lambda &= \langle a_0 \rangle_\lambda \langle b_0 \rangle_\lambda + \langle a_0 \rangle_\lambda \langle b_1 \rangle_\lambda + \langle a_1 \rangle_\lambda \langle b_0 \rangle_\lambda - \langle a_1 \rangle_\lambda \langle b_1 \rangle_\lambda \\ &= \langle a_0 \rangle_\lambda (\langle b_0 \rangle_\lambda + \langle b_1 \rangle_\lambda) + \langle a_1 \rangle_\lambda (\langle b_0 \rangle_\lambda - \langle b_1 \rangle_\lambda). \end{aligned} \tag{7.12}$$

Since every expectation value is in the range $[-1, 1]$, the last expression can be upper bounded by:

$$S_{\text{CHSH}}^\lambda \leq |\langle b_0 \rangle_\lambda + \langle b_1 \rangle_\lambda| + |\langle b_0 \rangle_\lambda - \langle b_1 \rangle_\lambda|. \tag{7.13}$$

Without loss of generality we can assume that: $\langle b_0 \rangle_\lambda \geq \langle b_1 \rangle_\lambda \geq 0$ (the other cases lead to the same result), which substituted in the last expression yields:

$$S_{\text{CHSH}}^\lambda \leq 2\langle b_0 \rangle_\lambda \leq 2. \tag{7.14}$$

By employing (7.14) in (7.11), we prove the CHSH inequality in (7.9) for every probability distribution that can be written in the form (7.6).

In order to complete the proof of Bell's theorem, we demonstrate that quantum theory predicts correlations violating the CHSH inequality (7.9) for a specific implementation of the Bell experiment. This implies that they cannot be explained in terms of an LHV model (7.6).

Suppose that Alice's system and Bob's system are qubits in the entangled (Bell) state:

$$|\Phi^+\rangle = \frac{1}{\sqrt{2}} (|00\rangle + |11\rangle), \tag{7.15}$$

whose corresponding density operator can be written in terms of the Pauli operators X, Y and Z as follows [24, 25]:

$$|\Phi^+\rangle\langle\Phi^+| = \frac{1}{4} (\mathbb{1} \otimes \mathbb{1} + X \otimes X + Z \otimes Z - Y \otimes Y). \tag{7.16}$$

Notably, the last expression decomposes the projector on the state $|\Phi^+\rangle$ as a sum over all the *stabilizer operators* of the state $|\Phi^+\rangle$. An operator O is a stabilizer of a state $|\psi\rangle$ if $O|\psi\rangle = |\psi\rangle$, i.e. if the state is an eigenstate of O with eigenvalue one.

In this setting, we describe Alice's (Bob's) measurement on her (his) qubit as a binary projective measurement represented by the observable A_x (B_y), corresponding to input x (y). The generic form of A_x (B_y) is given by $A_x = \boldsymbol{\alpha}^{(x)} \cdot \boldsymbol{\sigma}$ ($B_y = \boldsymbol{\beta}^{(y)} \cdot \boldsymbol{\sigma}$), where $\boldsymbol{\sigma}$ is the vector of Pauli operators: $\sigma_1 = X$, $\sigma_2 = Y$ and $\sigma_3 = Z$ and where $\|\boldsymbol{\alpha}^{(x)}\| = \|\boldsymbol{\beta}^{(y)}\| = 1$.

Then, the correlators in the CHSH inequality (7.9) can be written as:

$$\langle a_x b_y \rangle = \langle \Phi^+ | A_x \otimes B_y | \Phi^+ \rangle = \mathrm{Tr}\left[|\Phi^+\rangle\langle\Phi^+|(A_x \otimes B_y) \right]$$
$$= \alpha_1^{(x)}\beta_1^{(y)} - \alpha_2^{(x)}\beta_2^{(y)} + \alpha_3^{(x)}\beta_3^{(y)}, \tag{7.17}$$

where we used the decomposition (7.16), the multiplication rule of Pauli operators in (2.20) and the fact that the Pauli operators are traceless.

We aim at maximizing the CHSH value (7.8), expressed in terms of the measurement directions $\boldsymbol{\alpha}^{(x)}$ and $\boldsymbol{\beta}^{(y)}$ of Alice and Bob via (7.17):

$$S_{\mathrm{CHSH}} = \alpha_1^{(0)}(\beta_1^{(0)} + \beta_1^{(1)}) - \alpha_2^{(0)}(\beta_2^{(0)} + \beta_2^{(1)}) + \alpha_3^{(0)}(\beta_3^{(0)} + \beta_3^{(1)})$$
$$\alpha_1^{(1)}(\beta_1^{(0)} - \beta_1^{(1)}) - \alpha_2^{(1)}(\beta_2^{(0)} - \beta_2^{(1)}) + \alpha_3^{(1)}(\beta_3^{(0)} - \beta_3^{(1)}). \tag{7.18}$$

The last expression is maximized if we choose, for instance,

$$\boldsymbol{\alpha}^{(0)} = (1, 0, 0) \quad \boldsymbol{\alpha}^{(1)} = (0, 0, 1)$$
$$\boldsymbol{\beta}^{(0)} = \left(\frac{1}{\sqrt{2}}, 0, \frac{1}{\sqrt{2}} \right) \quad \boldsymbol{\beta}^{(1)} = \left(\frac{1}{\sqrt{2}}, 0, -\frac{1}{\sqrt{2}} \right), \tag{7.19}$$

With these measurement settings the CHSH value (7.18) is given by:

$$S_{\mathrm{CHSH}} = 2\sqrt{2} > 2, \tag{7.20}$$

i.e. the CHSH inequality (7.9) is violated.

Note that the measurement settings in (7.19) correspond to Alice and Bob measuring the observables:

$$A_0 = X \quad A_1 = Z$$
$$B_0 = \frac{X + Z}{\sqrt{2}} \quad B_1 = \frac{X - Z}{\sqrt{2}}, \tag{7.21}$$

which substituted into the CHSH expression (7.8) and upon simplifications lead to:

$$S_{\mathrm{CHSH}} = \sqrt{2}\langle \Phi^+ | X \otimes X | \Phi^+ \rangle + \sqrt{2}\langle \Phi^+ | Z \otimes Z | \Phi^+ \rangle = 2\sqrt{2}. \tag{7.22}$$

The last expression has the merit to show that it is optimal for the parties to choose measurements such that the resulting CHSH expression, after being simplified, is exclusively composed of correlators (expectation values) of stabilizers of the state $|\Phi^+\rangle$. This makes sense, since by definition the correlator of a stabilizer evaluated on the stabilized state achieves the maximum value of 1.

The CHSH violation predicted by quantum theory has profound consequences, as it implies that one of the three assumptions (7.3), (7.4) and (7.5) that led to the derivation of the CHSH inequality does not hold for quantum theory. The quantum theory we consider is the standard non-relativistic quantum theory, which describes quantum systems and measurements in terms of tensor-product Hilbert spaces and local Kraus operators acting on the corresponding Hilbert space. These features, combined with the partial trace rule, ensure that the local statistics of a system only depend on its reduced density operator. Therefore, no superluminal communication is allowed between parties and in particular the conditions of parameter independence (7.4) and of free will (7.5) are satisfied.

From the above argument, we conclude that quantum mechanics is a non-local theory, i.e. it does not satisfy the locality assumption in (7.3), and its predictions cannot be reproduced by any local theory. This concludes the proof of Bell's theorem.

Another important consequence of Bell's theorem are Bell inequalities. These can seen as means to experimentally test if nature behaves according to local theories or not. The numerous experiments demonstrating the violation of Bell inequalities have proved beyond reasonable doubt that nature is non-local. In particular, we mention the first successful results in this direction by Aspect et al. [26]. Recent and more sophisticated experiments have confirmed the existence of non-local correlations in *loophole-free* Bell tests [27–29], i.e. conducted without making any assumption that could lead to a description of the non-local correlation through an LHV model.

7.2 Local, Quantum, No-Signaling and Causal Correlations

In this Section we wish to clarify the relations existing between different types of correlations, starting from what we have seen in the proof of Bell's theorem.

First of all, we can derive the so called *no-signaling constraints* [21, 22] from the parameter-independence (7.4) and free-will (7.5) assumptions used in Bell's theorem:

$$
\begin{aligned}
p(a|x, y) &= \int d\lambda \, p(a|x, y, \lambda) p(\lambda|x, y) \\
&\overset{(7.5)}{=} \int d\lambda \, p(a|x, y, \lambda) p(\lambda) \\
&\overset{(7.4)}{=} \int d\lambda \, p(a|x, \lambda) p(\lambda) \\
&= p(a|x),
\end{aligned} \tag{7.23}
$$

and similarly one gets

$$p(b|x, y) = p(b|y). \tag{7.24}$$

Note that another way to write the no-signaling constraint in (7.23) is $\sum_b p(a, b|x, y) = \sum_b p(a, b|x, y')$ for every a, x, y and y'. The no-signaling constraints state that the probability distribution of the outcomes of one party is independent of the inputs of the other party.

The justification of the no-signaling constraints for spacelike separated parties comes from the *causality constraint* of special relativity, according to which one party cannot communicate with another party by sending a superluminal signal, i.e. a signal that travels faster than the speed of light. Indeed, the constraints on the probability distributions of a two-party Bell scenario imposed by causality coincide with (7.23) and (7.24).

The no-signaling constraints are generalized as follows in an N-party Bell scenario [30]:

$$\sum_{a_j} p(a_1, \ldots, a_j, \ldots, a_N | x_1, \ldots, x_j, \ldots, x_N)$$
$$= \sum_{a_j} p(a_1, \ldots, a_j, \ldots, a_N | x_1, \ldots, x_j', \ldots, x_N) \tag{7.25}$$
$$\forall j \in \{1, \ldots, N\}, \{a_1, \ldots, a_N\} \setminus \{a_j\}, \{x_1, \ldots, x_j, x_j', \ldots, x_N\},$$

stating that the probability distribution of the outcomes of any subset of parties is independent of the inputs of the complementary set of parties. Note that the constraints in (7.25) explicitly express this statement only for subsets of one party each, but the general statement can be deduced from (7.25) [22].

Interestingly, the multiparty no-signaling constraints (7.25) do not precisely capture the causality constraints for certain configurations of parties in the Minkowski spacetime (when $N \geq 3$). In other words, one can arrange the parties in spacetime such that the constraints on their probability distributions purely derived from causality are a strict subset of the no-signaling constraints in (7.25) [22]. By labelling $\mathcal{NS}$ the set of all possible correlations obeying the no-signaling constraints (7.25) and analogously $\mathcal{R}$ the set of correlations obeying the causality constraints of relativity, we have that: $\mathcal{NS} \subset \mathcal{R}$. We remark that this situation occurs only when the Bell scenario is composed of $N \geq 3$ parties, while for $N = 2$ we have that $\mathcal{NS} \equiv \mathcal{R}$ as stated above.

Considering again the bipartite Bell scenario, the set $\mathcal{Q}$ of quantum correlations is defined by those probability distributions that can be expressed as:

$$p(a, b|x, y) = \mathrm{Tr}\left[\rho_{AB} M_{a|x} \otimes M_{b|y}\right], \tag{7.26}$$

where ρ_{AB} is a quantum state on the joint Hilbert space $\mathcal{H}_A \otimes \mathcal{H}_B$ and $M_{a|x}$, $M_{b|y}$ are POVM elements (c.f. Sect. 2.2) relative to outcomes a, b given the inputs x, y.

Finally, the set $\mathcal{L}$ of local correlations is characterized by probability distributions that can be expressed in terms of an LHV model (7.6).

It is proved that every local correlation is also a quantum correlation, and that every quantum correlation satisfies the no-signaling constraints [21], as mentioned earlier. However, with the violation of the CHSH inequality (7.9), we have seen that there are quantum correlations outside the set of local correlations. Moreover, there are no-signaling correlations which are not quantum correlations. For instance, there are no-signaling correlations whose CHSH value S_{CHSH} achieves the algebraic bound of the expression: $S_{\mathrm{CHSH}} = 4$. Conversely, it is shown [21] that any quantum correlation leads to a CHSH value upper bounded by $2\sqrt{2}$, which is called the *Tsirelson bound*.

Due to these observations, the following strict inclusions hold: $\mathcal{L} \subset \mathcal{Q} \subset \mathcal{NS}$.

7.3 From Bell Violation to Security

Whenever a set of probability distributions, e.g., $p(a, b|x, y)$ in the bipartite Bell scenario, violates a Bell inequality, we talk about *Bell violation* and we call the correlations generated by such distributions *non-local*. In this Section we clarify the connection between Bell violation and the security of device-independent (DI) quantum cryptographic protocols.

The security of DI protocols, such as DIQKD and DIRG, is guaranteed irrespective of the trustworthiness of the devices used in their implementation. In a DI protocol, each party holds a device modelled as black box producing an output upon receiving an input from the party. By repeating this operation for several rounds, the parties collect a series of outcomes, each related to the input that generated it.

A fraction of the collected outputs forms the secret key shared by the parties in DIQKD and DICKA protocols, or the secret random bitstring in DIRG protocols. The remaining outputs are used to test a Bell inequality with a Bell experiment like the one described in Sect. 7.1.

Performing a Bell test during the execution of a device-independent (DI) protocol is crucial to ensure its security. Indeed, upon observing a Bell violation, the parties can certify that the random outcomes collected during the execution of the protocol are (at least partially) secret, i.e. unknown to a potential eavesdropper (Eve). What is the link between Bell violation and the privacy of the parties' outcomes?

Firstly, observing a Bell violation rules out the possibility that the outcomes collected by the parties have been generated by an LHV strategy (7.6). In particular, this excludes the possibility that the outcomes have been predetermined by Eve by setting up the systems such that the probabilities $p(a|x, \lambda)$ and $p(b|y, \lambda)$ are deterministic functions of x, y and λ: $p(a|x, \lambda), p(b|y, \lambda) \in \{0, 1\}$. This ensures that, even if the parties' systems were fabricated by Eve, she could not have predicted all the outcomes observed by the parties during the Bell experiment. This is a good starting point to have a secret string of random bits.

As we discussed in Bell's theorem proof (c.f. Sect. 7.1), quantum theory allows for correlations violating a Bell inequality. Specifically, a Bell violation occurs only when the parties share a quantum entangled state and their measurements are described by non-commuting observables (e.g., $[A_0, A_1] \neq 0$) [21]. One can thus interpret Bell inequalities—and violation thereof—as device-independent entanglement witnesses [3].

An important property of entanglement, called *monogamy of entanglement* [31, 32], states that if the quantum systems of two parties, say Alice and Bob, are strongly entangled, then a third quantum system shares little entanglement with them. Thanks to this property, upon observing a Bell violation, Alice and Bob are sure that Eve was poorly entangled with their systems and thus has little information on their outcomes. Hence the secrecy of the parties' outcomes is granted. Notably, the monogamy of correlations is not specific to quantum theory, rather it is present in any no-signaling theory leading to non-local correlations [21].

7.3.1 DIQKD Security Under Coherent Attacks

Let us now formalize the intuition provided above on the security of DI protocols. Here we focus on DIQKD protocols, but analogous considerations hold for DICKA and DIRG protocols.

The Bell violation estimated by the parties while running a DIQKD protocol only describes, on average, the amount of non-locality characterizing one protocol round, and in particular a round devoted to the generation of the secret key. This is enough to make a security statement for one round of the protocol. In particular, given the observed Bell violation, one can bound the conditional von Neumann entropy $H(R_A|E)$ of Alice's raw key bit R_A given Eve's side information E, which we already encountered when computing the secret key rates of QKD protocols (c.f. Chap. 3). The quantitative trade-off between Bell violation and conditional entropy is illustrated in Sect. 7.5 for the simplest DIQKD protocol.

However, the validity of the security statement for one round cannot be directly extended to the whole DI protocol and to all its outputs. The reason is that we consider the most general scenario where Eve performs coherent attacks, meaning that she can act differently in the various rounds and so can the devices. In fact, the security of DIQKD follows the same definitions and results reported in Sect. 3.3 for the security of general QKD schemes, where the main quantity to be estimated is the smooth min-entropy $H_{\min}^{\varepsilon}(R_A^n|E)$ of Alice's bits R_A^n given the side information available to Eve (see (3.27)). Here, Eve's side information includes her quantum system correlated with the initial state distributed to the parties' devices, but also the additional side information generated by the untrusted devices during the process.

In standard QKD,[2] we have discussed the existence of methods —such as the postselection technique (PST)—which reduce the security proof against coherent attacks to one against collective attacks, i.e. when the behaviour of the devices and the state distributed by Eve is the same in every round (c.f. Sect. 3.3.3). In this case,

---

[2] In this context, standard QKD refers to non-DI QKD.

one can focus on proving the security of one protocol round by using the asymptotic equipartition property (AEP) (c.f. (2.63)), which links the min-entropy of an i.i.d. state to the von Neumann entropy of one of its copies.

However, the methods used in standard QKD are not applicable to DIQKD. Recall, for instance, that the PST requires the knowledge of the Hilbert space dimension of the parties' systems, which is clearly not known in DIQKD.

Nevertheless, an important result named *entropy accumulation theorem* (EAT) [6, 33, 34] allows us to link the security of the whole DIQKD scheme to the security of one round and can be seen as a generalization of the AEP valid for non-i.i.d. rounds. In particular, the protocol rounds considered by EAT are such that the key bit $R_A^{(i)}$ generated in the i-th round can also depend on what happened in all the previous rounds, but not on the future rounds, which is a meaningful assumption in sequential DIQKD protocols. According to EAT, the amount of entropy accumulated during the described sequential processes, i.e. the smooth min-entropy $H_{\min}^\varepsilon(R_A^n|E)$, is at least n times the conditional von Neumann entropy of one round $H(R_A|E)$ evaluated over the observed Bell violation (up to correction factors of order $\sqrt{n}$).

Therefore, our discussion will now focus on quantitatively connecting the conditional von Neumann entropy of one protocol round with the Bell violation observed in the Bell test. In the next two Sections we explore this relationship in the context of the simplest example of a DIQKD protocol.

Finally, we remark that these considerations similarly hold for DIRG protocols, where a secret random bitstring is extracted from the collected outcomes of one party or more parties, who can be co-located—e.g., located in the same laboratory.

7.4 Device-Independent QKD

In this Section we summarize the assumptions that still hold in any bipartite DI protocol (the generalization to more parties is straightforward). We then illustrate the most common DIQKD protocol, which is based on the violation of the CHSH inequality [23]. In the next Section we prove the protocol's security by deriving a lower bound on the conditional von Neumann entropy of one round as a function of the observed CHSH violation.

7.4.1 Assumptions

Despite the fact that in a DI scenario no assumption is made on the quantum state shared by the parties, nor on its dimension and measurement, there are still some unavoidable assumptions in place [35]. Here we list them:

Isolated laboratories   No information flows in or out Alice's and Bob's labs except for what is established by the protocol, i.e. the state distribution in each round and the public classical communication between Alice and Bob.

Isolated source   The preparation of the states is independent of the measurements performed on them.

Trusted classical post-processing   The classical communication is performed over a public authenticated channel and the data is processed with trusted computers.

Trusted random number generators   Alice and Bob independently possess a trusted random number generator whose outcomes are only known to the owner.

Note that the complete removal of any of the above assumptions would lead to a strategy where the key is leaked to Eve [35].

7.4.2 DIQKD Based on the CHSH Inequality

Consider the following DIQKD protocol whose security is based on testing the CHSH inequality [3, 6]. Alice holds an uncharacterised device with two inputs $x \in \{0, 1\}$ and two outputs for each input: $a_x \in \{-1, 1\}$. Ideally, upon receiving an input, the device performs a measurement on Alice's portion of an entangled state that she shares with Bob and provides the outcome of the measurement. We emphasize, however, that we do not specify the implementation when proving the protocol's security. Similarly, Bob holds a device with three inputs $y \in \{0, 1, 2\}$ and two outputs per input $b_y \in \{-1, 1\}$.

Before initiating the protocol, Alice and Bob agree on a set of parameters: the total number of rounds M, the probability $p_t \in (0, 1)$ with which they perform a test round, the expected CHSH value $S_{\exp} \in (2, 2\sqrt{2}]$ and its tolerated statistical fluctuation $\delta \in (0, 2\sqrt{2} - 2)$.

The protocol comprises the following steps[3] [36]:

1. Alice and Bob perform a test round with probability p_t or a key-generation (KG) round with probability $1 - p_t$. The information on which round to perform can be provided to them by a short preshared key (c.f. Remark 4.2). The total number of rounds is M.

2. In a test round Alice (Bob) randomly selects an input $x \in \{0, 1\}$ ($y \in \{0, 1\}$) on her (his) device and collects the output a_x (b_y), i.e. the parties test the CHSH inequality. In a KG round, Alice (Bob) selects the predefined input $x = 1$ ($y = 2$) and records the output—her (his) raw key bit—in the random variable R_A (R_B).

3. In parameter estimation (PE) the parties reveal the inputs and outputs of every test round to compute the observed CHSH value S:

---

[3]We remark that there exist more sophisticated versions of the same scheme with an improved secret key rate [6]. However, since we are not interested in investigating the protocol's performance, we consider this simplified version.

$$S = \langle a_0 b_0 \rangle + \langle a_0 b_1 \rangle + \langle a_1 b_0 \rangle - \langle a_1 b_1 \rangle. \tag{7.27}$$

If $S < S_{\text{exp}} - \delta$, the protocol aborts. The parties additionally reveal a fraction of the KG outcomes to estimate the QBER E_{AB}:

$$E_{AB} = \Pr[R_A \neq R_B] \tag{7.28}$$

4. The parties perform one-way error correction (EC): Alice discloses some information on her raw key by communicating it to Bob via the classical public channel. With the information received from Alice, Bob computes a guess of her raw key. If the EC scheme fails, the protocol aborts.
5. The parties distil two secret keys from their error-corrected raw keys by applying a privacy amplification (PA) procedure.

In an ideal implementation of the above scheme, the CHSH inequality is maximally violated, implying that Eve has no information on the generated secret key (Sect. 7.5). In order for this to happen, the parties can e.g.. share the pure Bell state $|\Phi^+\rangle$ (7.15) in each round of the protocol. The measurements of Alice and Bob in the test rounds are given by (7.21) and are the same used to maximally violate the CHSH inequality in the example of Sect. 7.1.

In the DIQKD protocol, Bob has an additional setting $y = 2$ that is only used for KG. In order for Bob to have his raw key bits R_B perfectly correlated with Alice's, he must measure the same observable $B_2 = Z$ that Alice measures in a KG round. Indeed, in a KG round Alice measures $A_1 = Z$ (according to (7.21)) and the outcomes of two local Z measurements on the Bell state $|\Phi^+\rangle$ are perfectly correlated.

As explained in Sect. 7.3, thanks to EAT the security proof of the described DIQKD protocol in the finite-key scenario is reduced to the computation of the conditional von Neumann entropy $H(R_A|E)$, relative to one protocol round, as a function of the observed CHSH violation S. Note that, since without violation ($S \leq 2$) the estimation of the entropy $H(R_A|E)$ would be $H(R_A|E) = 0$ (see Sect. 7.5), from now one we refer to S as the CHSH *violation* rather than the CHSH *value*.

We point out that the security of the protocol is composable in the sense of the definition given in Sect. 3.3. However this is true only as far as the devices are not reused in another run of the protocol [35, 36].

In the asymptotic limit ($M \to \infty$) the finite-key effects become negligible and the asymptotic secret key rate constitutes an upper bound on the secret key rate achieved with finite resources [37]. The asymptotic secret key rate of the described DIQKD protocol coincides with the one of standard QKD protocols (3.9) and reads:

$$r = H(R_A|E) - H(R_A|R_B). \tag{7.29}$$

The second term in (7.29) is due to the classical information leaked during EC and can be estimated analogously to standard QKD in terms of the QBER E_{AB} (see (3.25)). However, differently from standard QKD schemes, here the entropy $H(R_A|E)$ is estimated device-independently as a function of the observed CHSH violation. This is the content of the next Section.

In a similar fashion, the asymptotic rate of secret bits generated by a DIRG protocol reads:

$$r = H(R_A|E), \tag{7.30}$$

where the term due to EC is removed since the only goal is to produce a secret random bitstring in one specific location.

7.5   Conditional Entropy Bound

We derive an analytical lower bound on the conditional von Neumann entropy $H(R_A|E)$, relative to the DIQKD protocol of Sect. 7.4, for a given CHSH violation S. This result yields a lower bound on the protocol's secret key rate both in the finite-key and asymptotic regimes.

The analytical lower bound on $H(R_A|E)$ was first derived in [3]. This fundamental result allows for analytical expressions of the secret key rates (secret randomness generation rates) of all the DIQKD (DIRG) protocols based on the CHSH inequality or reducible to a CHSH violation (e.g., the DICKA protocol in [9]). Indeed, there is no analytical DIQKD key rate which does not rely on the bound derived in [3].

There are other ways to lower bound $H(R_A|E)$ in terms of the violation of a given Bell inequality, which are employed when an analytical lower bound is not available. A common procedure is to numerically compute the min-entropy $H_{min}(R_A|E)$ [38–41] and use the fact that the min-entropy is a lower bound of the von Neumann entropy (see Eq. 2.56). However the bounds derived in this way are fairly loose, leading to poorly-performing DIQKD schemes.

The critical result derived in [3] is the reduction of the state shared by Alice and Bob in one round of the protocol to a two-qubit state which is diagonal in the Bell basis (3.15). Note that this result is derived assuming i.i.d. rounds in the DIQKD protocol above, i.e. Eve performs collective attacks. Nevertheless, as we discussed, the result can be applied to proving the protocol's security in the most general scenario thanks to EAT.

Theorem 7.1 *([3]). Let Alice and Bob perform the DIQKD protocol described in Sect. 7.4.2. It is not restrictive to assume that, in each round, Eve distributes a mixture $\sum_\alpha p_\alpha \rho_\alpha$ of two-qubit states ρ_α, together with a flag $|\alpha\rangle$ (known to her) which determines the measurements performed on ρ_α given the parties' inputs. Without loss of generality, the measurements performed by Alice's and Bob's devices on ρ_α are rank-one binary projective measurements in the (x, y)-plane of the Bloch sphere. Moreover, each state ρ_α is diagonal in the Bell basis (3.15) and reads:*

$$\rho_\alpha = \sum_{i,j=0}^{1} \lambda_{ij}^\alpha |\psi_{ij}\rangle\langle\psi_{ij}| \quad \text{with} \quad \lambda_{0j}^\alpha \geq \lambda_{1j}^\alpha \; \forall j \in \{0, 1\}. \tag{7.31}$$

The proof of Theorem 7.1 is given in the Appendix of this chapter (Sect. 7.8) and is rearranged in order to coherently fit with the more general result proved in [42].

The second crucial ingredient to derive the bound on $H(R_A|E)$ is the analytical expression of the maximal CHSH violation $\mathcal{S}_\rho$ that can achieved on a given two-qubit state ρ. In other words, there exist measurements performed by Alice and Bob on ρ such that the observed CHSH violation is $S = \mathcal{S}_\rho$, and any other measurement setting leads to violations $S \leq \mathcal{S}_\rho$. This is a well-known result derived in [43].

Theorem 7.2 ([43]) *The maximum violation $\mathcal{S}_\rho$ of the CHSH inequality (7.27), attained by a two-qubit state ρ, is given by:*

$$\mathcal{S}_\rho = 2\sqrt{t_0 + t_1} \tag{7.32}$$

where t_0 and t_1 are the largest and second-to-the-largest eigenvalues of the matrix $T_\rho T_\rho^T$, where T_ρ is the correlation matrix of ρ, with elements: $[T_\rho]_{ij} = \text{Tr}[\rho(\sigma_i \otimes \sigma_j)]$ for $i, j = 1, 2, 3$ (σ_i are the Pauli matrices).

For the state ρ_α in (7.31), the maximal CHSH violation reads:

$$\mathcal{S}_\alpha = 2\sqrt{2} \max\left\{\sqrt{(\lambda_{00}^\alpha - \lambda_{11}^\alpha)^2 + (\lambda_{01}^\alpha - \lambda_{10}^\alpha)^2}, \sqrt{(\lambda_{00}^\alpha - \lambda_{10}^\alpha)^2 + (\lambda_{01}^\alpha - \lambda_{11}^\alpha)^2}\right\}.$$
$$\tag{7.33}$$

We are now ready to derive the lower bound on the conditional entropy $H(R_A|E)$ in terms of the observed CHSH violation S. The derivation provided here, although based on the same concepts used in [3], presents further details in order to better guide the reader in the various steps. Moreover, the last step of the proof leading to the final result substantially differs from [3] as it employs a completely different approach.

To start with, Theorem 7.1 says that we can restrict the computation of the conditional entropy of interest over a mixture of states ρ_α of the form (7.31). We emphasize that the total information available to Eve includes the knowledge of the value α stored in a random variable Ξ, therefore we must compute the conditional entropy $H(R_A|E_{\text{tot}})$, where $E_{\text{tot}} = E\Xi$.

More specifically, the aim is to derive a lower bound $F(S)$ of $H(R_A|E_{\text{tot}})$ where F is a function of the observed CHSH violation S. The bound is tight if for every violation S there exist a quantum state and a set of measurements that achieve that violation and whose conditional entropy is exactly given by $F(S)$.

By the argument above,[4] we can be express the conditional entropy $H(R_A|E_{\text{tot}})$ as follows:

$$H(R_A|E_{\text{tot}}) = \sum_\alpha p_\alpha H(R_A|E \Xi = \alpha)$$

$$= \sum_\alpha p_\alpha H(R_A|E)_{\rho_\alpha}, \tag{7.34}$$

where $H(R_A|E)_{\rho_\alpha}$ is the conditional entropy of Alice's raw key bit given that Eve distributed the state ρ_α. Similarly, the observed violation S can be written as:

$$S = \sum_\alpha p_\alpha S_\alpha, \tag{7.35}$$

where S_α is the violation that the parties would observe if they were given the state ρ_α in each round.

We can then focus on deriving a lower bound on $H(X|E)_{\rho_\alpha}$:

$$H(R_A|E)_{\rho_\alpha} \geq F(S_\alpha), \tag{7.36}$$

where F is a convex function of the violation S_α. Indeed, by combining (7.34), (7.35), (7.36) and the convexity of F, we get the desired lower bound on $H(R_A|E_{\text{tot}})$ as a function of the observed violation S:

$$H(R_A|E_{\text{tot}}) \geq F(S). \tag{7.37}$$

Remark 7.1 The task is reduced to minimizing the conditional entropy $H(R_A|E)_{\rho_\alpha}$ over all the states ρ_α of the form (7.31) (Theorem 7.1), whose CHSH violation S_α is upper bounded by (7.33) (Theorem 7.2). In doing so, we obtain an explicit expression for $F(S_\alpha)$ in (7.36).

We start by providing Eve with the maximum amount of side information (as in every QKD protocol) by assuming that the state on $\mathcal{H}_A \otimes \mathcal{H}_B \otimes \mathcal{H}_E$ is pure, i.e. Eve holds the purifying system of ρ_α. Considering that ρ_α is written in its spectral decomposition in (7.31), we have the following pure state on $\mathcal{H}_A \otimes \mathcal{H}_B \otimes \mathcal{H}_E$:

$$|\phi_{ABE}^\alpha\rangle = \sum_{i,j=0}^{1} \sqrt{\lambda_{ij}^\alpha} |\psi_{ij}\rangle \otimes |e_{ij}\rangle, \tag{7.38}$$

where $\{|e_{ij}\rangle\}_{i,j=0}^1$ is an orthonormal basis in $\mathcal{H}_E$.

---

[4] As a matter of fact, the quantum state on which $H(R_A|E_{\text{tot}})$ is computed is a c.q. state derived from (7.76), which is given in the proof of Theorem 7.1. Recall the formula to compute the conditional entropy of c.q. states: (2.52).

We then decompose the entropy $H(R_A|E)_{\rho_\alpha}$ according to the definition of conditional von Neumann entropy:

$$H(R_A|E)_{\rho_\alpha} = H(E|R_A)_{\rho_\alpha} + H(R_A)_{\rho_\alpha} - H(E)_{\rho_\alpha}. \tag{7.39}$$

Due to the fact that the state on $\mathcal{H}_A \otimes \mathcal{H}_B \otimes \mathcal{H}_E$ is pure, we can directly compute $H(E)_{\rho_\alpha}$ as follows (see Schmidt decomposition, Sect. 2.4):

$$H(E)_{\rho_\alpha} = H(AB)_{\rho_\alpha} = H(\{\lambda_{ij}^\alpha\}), \tag{7.40}$$

where the entropy on the r.h.s. is the Shannon entropy of the probability distribution defined by the eigenvalues λ_{ij}^α of ρ_α. Indeed, the eigenvalues of a density operator sum to one and are non-negative, due to the normalization and positivity of the density operator.

Note that the entropies in (7.39) are computed on the quantum state $\rho_{R_A E}^\alpha$ obtained by applying Alice's projective measurement corresponding to input $x = 1$ (the input for KG) on the pure state $|\phi_{ABE}^\alpha\rangle$ and by tracing out Bob's system. According to Theorem 7.1, Alice's measurement is described by a quantum operation $\mathcal{E}_{R_A}$ which projects on the eigenstates $\{|a\rangle\}_{a=0}^1$ of a generic observable in the (x, y)-plane: $A = \cos(\varphi)X + \sin(\varphi)Y$, with $\varphi \in [0, 2\pi]$. The eigenstates of A are given by:

$$|a\rangle_{R_A} = \frac{1}{\sqrt{2}}(|0\rangle + (-1)^a e^{i\varphi}|1\rangle), \tag{7.41}$$

and the corresponding measurement outcomes are defined as $a = 0, 1$ ($a = 0$ corresponds to eigenvalue $+1$ and $a = 1$ to eigenvalue -1). The state $\rho_{R_A E}^\alpha$ thus reads:

$$\rho_{R_A E}^\alpha \overset{(7.38)}{=} (\mathcal{E}_{R_A} \otimes \mathbb{1}_E) \, \mathrm{Tr}_B \left[|\phi_{ABE}^\alpha\rangle\langle\phi_{ABE}^\alpha| \right]$$

$$= \sum_{a=0,1} |a\rangle\langle a|_{R_A} \otimes \sum_{i,j,k,l=0,1} \sqrt{\lambda_{ij}^\alpha \lambda_{kl}^\alpha} \, \mathrm{Tr}_B[\langle a|\psi_{ij}\rangle\langle\psi_{kl}|a\rangle] |e_{ij}\rangle\langle e_{kl}|_E$$

$$=: \frac{1}{2} \sum_{a=0,1} |a\rangle\langle a|_{R_A} \otimes \rho_E^{\alpha,a}, \tag{7.42}$$

where we defined the normalized conditional state of Eve $\rho_E^{\alpha,a}$, given that Alice's raw key bit is equal to a.

The matrix representing $\rho_E^{\alpha,a}$ in the orthonormal basis $|e_{ij}\rangle$ is given by the following Hermitian matrix[5]:

---

[5] The missing entries are fixed by the fact that the matrix is Hermitian.

$$
\rho_E^{\alpha,a} =
\begin{bmatrix}
\lambda_{00}^\alpha & 0 & (-1)^a \sqrt{\lambda_{00}^\alpha \lambda_{01}^\alpha} \cos\varphi & (-1)^{a+1} \sqrt{\lambda_{00}^\alpha \lambda_{11}^\alpha} \, i \sin\varphi \\
\lambda_{10}^\alpha & (-1)^a \sqrt{\lambda_{10}^\alpha \lambda_{01}^\alpha} \, i \sin\varphi & (-1)^{a+1} \sqrt{\lambda_{10}^\alpha \lambda_{11}^\alpha} \cos\varphi \\
& \lambda_{01}^\alpha & 0 \\
& & \lambda_{11}^\alpha
\end{bmatrix},
\tag{7.43}
$$

with non-zero eigenvalues that are independent of a and given by:

$$
\eta_\pm(\varphi) = \frac{1}{2}\left[1 \pm \sqrt{(\lambda_{00}^\alpha - \lambda_{10}^\alpha)^2 + (\lambda_{01}^\alpha - \lambda_{11}^\alpha)^2 + 2(\lambda_{00}^\alpha - \lambda_{10}^\alpha)(\lambda_{01}^\alpha - \lambda_{11}^\alpha)\cos(2\varphi)}\right].
\tag{7.44}
$$

From (7.42), one immediately deduces that the reduced state on $\mathcal{H}_{R_A}$ is: $\rho_{R_A} = (1/2)\sum_{a=0,1}|a\rangle\langle a|$, hence its entropy is maximal:

$$
H(R_A)_{\rho_\alpha} = 1.
\tag{7.45}
$$

Moreover, by exploiting the fact that the state $\rho_{R_A E}^\alpha$ in (7.42) is a c.q. state, we can recast the expression for the entropy $H(E|R_A)_{\rho_\alpha}$ as follows (c.f. (2.52)):

$$
H(E|R_A)_{\rho_\alpha} = \frac{1}{2}\left(H(\rho_E^{\alpha,0}) + H(\rho_E^{\alpha,1})\right).
\tag{7.46}
$$

Now, the von Neumann entropy of the states $\rho_E^{\alpha,a}$ (for $a = 0, 1$) is simply given by the Shannon entropy of their eigenvalues (7.44). The entropy $H(E|R_A)_{\rho_\alpha}$ is thus given by:

$$
H(E|R_A)_{\rho_\alpha} = h(\eta_+(\varphi)),
\tag{7.47}
$$

where we used the definition of binary entropy (2.46) and the fact that the eigenvalues of a quantum state sum to one.

Note that the entropy in (7.47) depends on the angle φ which determines the direction of Alice's KG measurement in the (x, y)-plane of the Bloch sphere. Since in a DI scenario we do not have any information on the measurement direction, we have to consider the worst-case scenario, i.e. the direction that minimizes Eve's uncertainty and thus $H(E|R_A)_{\rho_\alpha}$. The function in (7.47) is clearly minimized for $\varphi = 0$ and simplifies to:

$$
H(E|R_A)_{\rho_\alpha} = h(\lambda_{00}^\alpha + \lambda_{01}^\alpha),
\tag{7.48}
$$

By substituting the results (7.40), (7.45) and (7.48) into (7.39), we can express the entropy $H(R_A|E)_{\rho_\alpha}$ to be minimized as follows:

$$
H(R_A|E)_{\rho_\alpha} = 1 - H(\{\lambda_{ij}^\alpha\}) + h(\lambda_{00}^\alpha + \lambda_{01}^\alpha).
\tag{7.49}
$$

We can now formulate the optimization problem (stated in Remark 7.1) whose solution is the lower bound (7.36) on $H(R_A|E)_{\rho_\alpha}$. The optimization problem reads as follows:

$$F(S_\alpha) := \min_{\{\lambda_{ij}^\alpha\}} 1 - H(\{\lambda_{ij}^\alpha\}) + h(\lambda_{00}^\alpha + \lambda_{01}^\alpha)$$

$$\text{sub. to}\quad \mathcal{S}_\alpha \geq S_\alpha \; ; \; \lambda_{0j}^\alpha \geq \lambda_{1j}^\alpha \; ; \; \sum_{i,j=0,1} \lambda_{ij}^\alpha = 1, \qquad (7.50)$$

and its detailed solution is given in the Appendix of this chapter (Sect. 7.9). We remark that the solution presented in this book is based on a completely different approach with respect to the original derivation in [3]. In particular, the proposed solution is inspired by similar proofs contained in [42] where the results of this section are extended to multiparty scenarios.

The solution of the above optimization is given by:

$$F(S_\alpha) = 1 - h\left(\frac{1}{2} + \frac{1}{2}\sqrt{\left(\frac{S_\alpha}{2}\right)^2 - 1}\right), \qquad (7.51)$$

which is a convex function as required by (7.37). Hence, by employing (7.51) in (7.37), we finally obtain the lower bound on the conditional von Neumann entropy of Alice's raw key bit as a function of the observed CHSH violation S:

$$H(R_A|E_{\text{tot}}) \geq 1 - h\left(\frac{1}{2} + \frac{1}{2}\sqrt{\left(\frac{S}{2}\right)^2 - 1}\right). \qquad (7.52)$$

By employing the derived bound e.g., in the asymptotic secret key rate (7.29) of the described DIQKD protocol, one can obtain a lower bound on the achievable key rate in terms of the observed CHSH violation S. We stress the fact that the bound in (7.52) plays a crucial role in obtaining an analytical expression for the secret key rate of any DIQKD protocol based on the CHSH inequality.

7.5.1 Privacy Certification in Standard- and DI-QKD

The conditional entropy bound in (7.52) can be seen as a quantitative certification of the privacy of Alice's key bit, in the context of a DIQKD protocol based on the CHSH inequality. It is interesting to compare this result with the analogous privacy certification (conditional entropy bound) used in a standard QKD protocol, namely the BB84 protocol studied in Sect. 3.2.

In order to carry out a fair comparison, we set equal grounds for the DIQKD protocol and the BB84 protocol. In particular, we have seen that the ideal resource state distributed in each round to Alice and Bob is the Bell state $|\Phi^+\rangle$ for both protocols (see (7.15) and (3.7)). In a more realistic scenario, the pure state $|\Phi^+\rangle$ undergoes a depolarizing channel (c.f. Sect. 2.5.1) generating the following mixed state:

$$\rho_{AB} = q|\Phi^+\rangle\langle\Phi^+| + (1-q)\frac{\mathbb{1}_A \otimes \mathbb{1}_B}{4}. \qquad (7.53)$$

We thus assume that in both protocols the state in (7.53) is distributed to Alice and Bob in every round. Then, in the DIQKD protocol the observed Bell violation reads $S = 2\sqrt{2}q$ when the parties perform the measurements given in (7.21) which are optimal[6] for the Bell state $|\Phi^+\rangle$.

This leads to the following conditional entropy bound (7.52) for the DIQKD protocol:

$$H(R_A|E)_{\text{DIQKD}} = 1 - h\left(\frac{1}{2} + \frac{1}{2}\sqrt{2q^2 - 1}\right). \qquad (7.54)$$

In the entanglement-based BB84 protocol described in Sect. 3.2, Alice and Bob perform measurements in the Z basis for key generation and in the X basis to estimate Eve's knowledge. The QBER in the X basis, given that they share the state in (7.53), reads: $E_X = (1-q)/2$. This leads to the following conditional entropy bound (3.24) for the BB84 protocol:

$$H(R_A|E)_{\text{BB84}} = 1 - h\left(\frac{1-q}{2}\right). \qquad (7.55)$$

We emphasize that DIQKD removes most of the assumptions on the measurement devices that typically hold in a BB84 protocol, where the additional assumptions need to be verified experimentally. However, the price to pay is a reduced capability of certifying the privacy of Alice's bit compared to the BB84 protocol, given that the parties share the same quantum state.

This is clear from Fig. 7.2, where we plot the conditional entropy bound of the DIQKD protocol (7.54) and of the BB84 protocol (7.55) as a function of the mixing parameter q of the depolarizing channel. Indeed, in the BB84 protocol Eve's uncertainty on Alice's bit is non-zero as soon as a fraction of the shared state is an entangled state. Conversely, in the DIQKD protocol Eve's uncertainty is only certified in the presence of a CHSH violation, which requires a much larger fraction of entanglement in the shared state ($q > 1/\sqrt{2}$).

---

[6]Note that the maximally mixed state $\frac{\mathbb{1}_A \otimes \mathbb{1}_B}{4}$ in (7.53) does not contribute to the violation S.

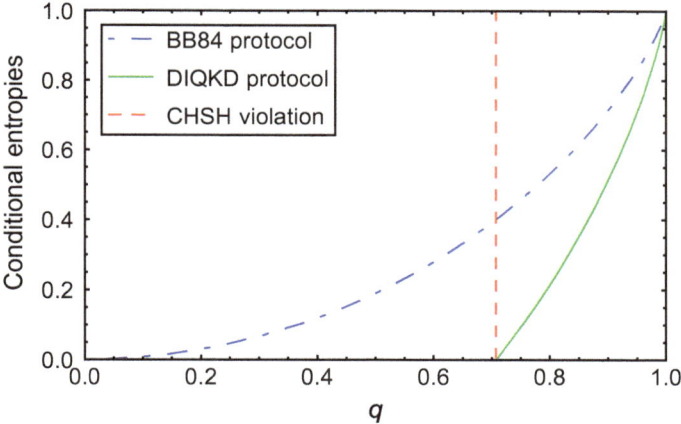

Fig. 7.2 Conditional von Neumann entropy $H(R_A|E)$ certified by a CHSH-based DIQKD protocol (solid green, Eq. 7.54) and by a BB84 protocol (dot-dashed blue, Eq. 7.55), as a function of the mixing parameter q of the depolarized Bell state (7.53) shared by Alice and Bob. We observe that, opposed to the BB84 protocol, the conditional entropy of the DIQKD protocol is non-zero only when Alice and Bob share a state that leads to a CHSH violation, i.e. when q is greater than the CHSH violation threshold (red dashed line)

7.6   Entropy Bounds for Multipartite Protocols

As anticipated in the previous Section, Theorems 7.1 and 7.2 have been generalized to multiparty DI scenarios in [42]. Such results allow for the derivation of accurate analytical bounds on conditional entropies of interest for multipartite DIRG protocols. We stress the fact that tighter bounds on the conditional entropies enhance the protocol's noise tolerance and relax the strict experimental requirements typical of DI protocols.

Consider a DI scenario with N parties that are denoted Alice$_1$, ..., Alice$_N$ for simplicity. In performing a DI protocol, the parties test a generic full-correlator Bell inequality [18, 44] with two dichotomic observables $A_x^{(i)}$ ($x = 0, 1$) per party ($i = 1, \ldots, N$). We call this an $(N, 2, 2)$ Bell scenario. A full-correlator Bell inequality is an inequality whose correlators always involve every party, i.e. they are of the form:

$$\langle A_{x_1}^{(1)} \cdots A_{x_N}^{(N)} \rangle. \tag{7.56}$$

From the observed Bell violation, the parties can certify the privacy of their outcomes by computing an appropriate conditional von Neumann entropy and thus determine the asymptotic rate of secret random bits generated by their DIRG or DICKA protocol.

In order to illustrate the generalized state reduction valid for an arbitrary $(N, 2, 2)$ Bell scenario, we first define the generalization of the Bell basis (3.15) in an N-qubit space [45].

Definition 7.1 The GHZ basis is composed of the following 2^N states:

$$|\psi_{\sigma,\mathbf{u}}\rangle = \frac{1}{\sqrt{2}}\left(|0\rangle|\mathbf{u}\rangle + (-1)^\sigma|1\rangle|\bar{\mathbf{u}}\rangle\right), \tag{7.57}$$

where $\sigma \in \{0,1\}$ while $\mathbf{u} \in \{0,1\}^{N-1}$ and $\bar{\mathbf{u}} = \mathbf{1} \oplus \mathbf{u}$ are $(N-1)$-bit strings.

We can now state the generalization of Theorem 7.1 to an $(N,2,2)$ Bell scenario.

Theorem 7.3 *([42]). Consider N parties testing an $(N,2,2)$ full-correlator Bell inequality. It is not restrictive to assume that, in each round, Eve distributes a mixture $\sum_\alpha p_\alpha \rho_\alpha$ of N-qubit states ρ_α, together with a flag $|\alpha\rangle$ (known to her) which determines the measurements performed on ρ_α given the parties' inputs. Without loss of generality, the measurements performed by each device on ρ_α are rank-one binary projective measurements in the (x,y)-plane of the Bloch sphere. Moreover, each state ρ_α is diagonal in the GHZ basis, except for some purely imaginary off-diagonal terms:*

$$\rho_\alpha = \sum_{\mathbf{u}\in\{0,1\}^{N-1}} \lambda_{0\mathbf{u}}^\alpha|\psi_{0,\mathbf{u}}\rangle\langle\psi_{0,\mathbf{u}}| + \lambda_{1\mathbf{u}}^\alpha|\psi_{1,\mathbf{u}}\rangle\langle\psi_{1,\mathbf{u}}| + is_{\mathbf{u}}^\alpha\left(|\psi_{0,\mathbf{u}}\rangle\langle\psi_{1,\mathbf{u}}| - |\psi_{1,\mathbf{u}}\rangle\langle\psi_{0,\mathbf{u}}|\right).$$

$$\tag{7.58}$$

Finally, N arbitrary off-diagonal terms $s_{\mathbf{u}}^\alpha$ can be assumed to be zero and the corresponding diagonal elements $(\lambda_{0\mathbf{u}}^\alpha, \lambda_{1\mathbf{u}}^\alpha)$ can be arbitrarily ordered (e.g., $\lambda_{0\mathbf{u}}^\alpha \geq \lambda_{1\mathbf{u}}^\alpha$).

We remark that Theorem 7.3 reduces to Theorem 7.1 when one considers the CHSH Bell scenario ($N=2$).

The second main tool to derive conditional entropy bounds is an analytical expression for the maximal violation of the considered Bell inequality achievable by a given state (e.g., Theorem 7.2 in Sect. 7.5). In [42] we derive such a result for a specific $(N,2,2)$ full-correlator inequality, namely the Mermin-Ardehali-Belinskii-Klyshko (MABK) inequality [46–48]. The MABK inequality is a multiparty generalization of the CHSH inequality and is obtained on the following MABK operator.

Definition 7.2 The MABK operator M_N is defined by recursion [49, 50]:

$$
\begin{aligned}
M_2 &= G_{\text{CHSH}}(A_0^{(1)}, A_1^{(1)}, A_0^{(2)}, A_1^{(2)}) \\
&\equiv A_0^{(1)} \otimes A_0^{(2)} + A_0^{(1)} \otimes A_1^{(2)} + A_1^{(1)} \otimes A_0^{(2)} - A_1^{(1)} \otimes A_1^{(2)} \\
M_N &= \frac{1}{2}G_{\text{CHSH}}(M_{N-1}, \overline{M_{N-1}}, A_0^{(N)}, A_1^{(N)}),
\end{aligned}
\tag{7.59}
$$

where $A_x^{(i)}$ ($i=0,1$) is the x-th binary observable of Alice$_i$ and where $\overline{M_l}$ is the operator obtained from M_l by replacing every observable $A_x^{(i)}$ with $A_{1-x}^{(i)}$.

Then the N-partite MABK inequality reads as follows:

$$\langle M_N \rangle = \mathrm{Tr}[M_N \rho] \leq \begin{cases} 2, & \text{classical bound} \\ 2^{N/2}, & \text{GME threshold} \\ 2^{(N+1)/2} & \text{quantum bound} \end{cases} \tag{7.60}$$

where M_N is the MABK operator and a violation of the GME threshold implies that the parties share a genuine multipartite entangled (GME) state (c.f. Definition 2.4).

We now present the upper bound on the maximal violation of the N-partite MABK inequality derived in [42]. This result can be seen as a generalization of Theorem 7.2 since the latter is recovered for $N = 2$.

Theorem 7.4 *([42]). The maximum violation M_ρ of the N-partite MABK inequality (7.60), attained by rank-one projective measurements on a given N-qubit state ρ, satisfies*

$$M_\rho \leq 2\sqrt{t_0 + t_1} \tag{7.61}$$

where t_0 and t_1 are the largest and second-to-the-largest eigenvalues of the matrix $T_\rho T_\rho^T$, where T_ρ is the correlation matrix of ρ.

The correlation matrix of an N-qubit state can be defined as follows.

Definition 7.3 The correlation matrix T_ρ of an N-qubit state ρ is defined by the matrix elements $[T_\rho]_{ij} = \mathrm{Tr}[\rho \sigma_{\nu_1} \otimes \cdots \otimes \sigma_{\nu_N}]$ such that:

$$i = 1 + \sum_{k=1}^{\lceil N/2 \rceil} 3^{\lceil N/2 \rceil - k}(\nu_k - 1)$$

$$j = 1 + \sum_{k=\lceil N/2 \rceil + 1}^{N} 3^{N-k}(\nu_k - 1) \tag{7.62}$$

where $\nu_1, \ldots, \nu_N \in \{1, 2, 3\}$, $\sigma_1 = X$, $\sigma_2 = Y$ and $\sigma_3 = Z$ are the Pauli matrices and $\lceil x \rceil$ returns the smallest integer greater or equal to x.

We remark that the upper bound on the maximal MABK violation in Theorem 7.4 is only tight on certain classes of states, differently from its bipartite counterpart, Theorem 7.2. Opposed to Theorem 7.2, Theorem 7.4 also restricts the measurements on each qubit to rank-one projective measurements (defined by combinations of Pauli operators) in agreement with the result of Theorem 7.3, thus excluding the identity as a viable observable. Note that for $N = 2$ the identity would not lead to any violation [43], hence Theorem 7.4 effectively reduces to Theorem 7.2.

To the best of our knowledge, Theorem 7.4 is the first result of such kind valid for an N-partite Bell inequality. Recently a similar bound was derived in the $N = 3$ case [51]. However, Theorem 7.4 is proved to be tight on a larger set of states and is valid for an arbitrary number of parties N.

7.6.1 Conditional Entropy Bounds for Three Parties

Equipped with the results of Theorems 7.3 and 7.4, we are able to obtain analytical bounds on conditional von Neumann entropies that are relevant for the security of certain multipartite DI protocols [42].

Specifically, we consider the $(3, 2, 2)$ Bell scenario depicted in Fig. 7.3 where Alice, Bob and Charlie test the tripartite MABK inequality in order to certify the privacy of some of their outcomes, by deriving lower bounds on suitable conditional von Neumann entropies. In particular, we obtain bounds on the conditional von Neumann entropies $H(R_A|E)$ and $H(R_A R_B|E)$ as a function of the observed MABK violation S. The bounds derivation is similar to the one described in Sect. 7.5 for two parties, although it presents additional difficulties due to the increased number of parties and outcomes [42].

We recall that the entropy $H(R_A|E)$ determines the asymptotic rate of secret random bits generated at Alice's location by a multiparty DIRG or DICKA protocol[7] [33, 52]. Similarly, the bound on the entropy $H(R_A R_B|E)$ can represent the rate at which co-located parties generate DI global randomness from Alice and Bob's outcomes [33, 53].

In Fig. 7.4 we plot the lower bounds on $H(R_A|E)$ and $H(R_A R_B|E)$ as a function of the observed MABK violation S. The analytical expressions corresponding to the plotted curves are given by [42]:

$$H(R_A|E) \geq 1 - h\left(\frac{1}{2} + \frac{1}{2}\sqrt{\frac{S^2}{8} - 1}\right) \tag{7.63}$$

$$H(R_A R_B|E) \geq 2 - H\left(\{1 - 3f(S), f(S), f(S), f(S)\}\right), \tag{7.64}$$

where $h(x)$ is the binary entropy (c.f. Eq. 2.46), $H(\{p_x\})$ is the Shannon entropy of the probability distribution p_x and $f(S)$ is defined as:

$$f(S) := \frac{1}{4} - \frac{\sqrt{3}}{24}\sqrt{S^2 - 4}. \tag{7.65}$$

Some comments are due. From Fig. 7.4 we observe that the lower bound on $H(R_A|E)$ is null for violations of the tripartite MABK inequality below the GME threshold. In [42] we additionally present a lower bound on $H(R_A|E)$ when an arbitrary number of parties N test the N-partite MABK inequality. Even the N-party bound is null for violations below the N-partite GME threshold.

Given that the bounds on $H(R_A|E)$ are tight at the GME threshold, we deduce that GME is necessary to certify the privacy of a party's outcome in any DI scenario based on the MABK inequality. Being the latter a prerequisite of any DICKA protocol (not necessarily based on the MABK inequality), it is an open question whether GME is

---

[7]In a DICKA protocol, the asymptotic conference key rate is given by $H(R_A|E)$ from which one subtracts the information leaked during error correction.

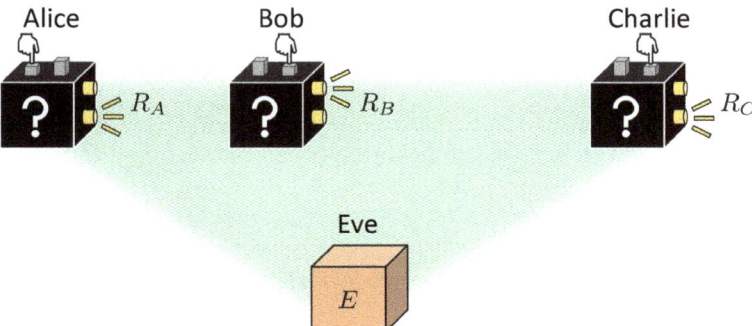

Fig. 7.3 Alice, Bob and Charlie generate randomness with the outputs R_A, R_B and R_C of their unknown quantum devices, whose privacy is certified by testing the MABK inequality. Each device is equipped with two inputs and two outputs. Eve might hold a quantum memory E entangled with the parties' devices and use it to guess the parties' outcomes. We compute Eve's uncertainty on Alice's outcome R_A by deriving an analytical expression for the conditional von Neumann entropy $H(R_A|E)$. We also assume that Alice and Bob are co-located and collaborate to generate global randomness from their outcomes R_A and R_B. In this case, we quantify Eve's uncertainty on their outcomes by computing $H(R_A R_B|E)$

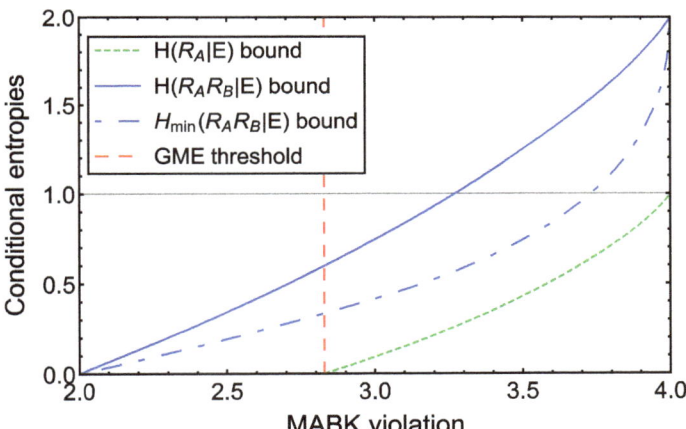

Fig. 7.4 Lower bounds on the conditional von Neumann entropies $H(R_A|E)$ (dotted green, Eq. 7.63) and $H(R_A R_B|E)$ (solid blue, Eq. 7.64) and on the conditional min-entropy $H_{\min}(R_A R_B|E)$ (dot-dashed blue, Eq. 7.66) as a function of the MABK violation observed by three parties. We notice that Eve has full information on Alice's outcome R_A for violations below the GME threshold (dashed red line). Moreover, bounding Eve's uncertainty on Alice and Bob's outcomes with the suitable von Neumann entropy (blue solid line) brings a substantial advantage compared to bounding the correspondent min-entropy (blue dot-dashed line)

necessary for a successful implementation of a DICKA protocol. Conversely, it has been shown that GME is not necessary to perform a device-dependent CKA protocol [54].

The lower bound on the von Neumann entropy $H(R_A R_B|E)$ is plotted in Fig. 7.4 together with a tight lower bound on the correspondent min-entropy[8] $H_{\min}(R_A R_B|E)$, given by [53]:

$$H_{\min}(R_A R_B|E) \geq -\log p^\uparrow_{\text{guess}}(R_A R_B|E) \tag{7.66}$$

where $p^\uparrow_{\text{guess}}(R_A R_B|E)$ is a tight upper bound on Eve's joint guessing probability of R_A and R_B:

$$p^\uparrow_{\text{guess}}(R_A R_B|E) = \begin{cases} \frac{3}{4} - \frac{S}{8} + \sqrt{3}\sqrt{\frac{S}{8}\left(\frac{1}{2} - \frac{S}{8}\right)} & \text{if } S > 3 \\ \frac{3}{2} - \frac{S}{4} & \text{if } S \leq 3. \end{cases} \tag{7.67}$$

Fig. 7.4 confirms that there is a significant improvement in certifying the privacy of Alice and Bob's outcomes, from a given MABK violation, with the bound on the von Neumann entropy (7.64), as opposed to using the more accessible min-entropy (7.66). This has a direct impact on the performance of DI (global) randomness generation protocols, since it increases the fraction of generated random bits proved to be secret.

The last observation demonstrates the potential of the analytical approach [42] in bounding the von Neumann entropies of interest in DI protocols. In particular, the developed techniques could pave the way for similar results valid for the Bell inequalities employed in the existing DICKA protocols [9, 10]. In this context, the derivation of tight analytical bounds on the von Neumann entropy would translate to tight security proofs—which are still missing—and to increased protocol performance.

7.7 Device-Independent Conference Key Agreement

In this Section we argue on the potential applicability of the conditional entropy bounds of Sect. 7.6 to the security proofs of DI conference key agreement (DICKA) protocols. In particular, we investigate the relationship between the structure of full-correlator Bell inequalities and the task of DICKA. We conclude the Chapter by providing an example of multipartite Bell inequality suited to DICKA protocols [10].

---

[8]We remark that the min-entropy is often used to lower bound the von Neumann entropy in DI protocols, since it can be directly estimated from the observed statistics of the Bell test [38–41] and since Eq. (2.56) holds.

7.7.1 Full-Correlator Bell Inequalities and DICKA

All the results presented in Sect. 7.6 stem from the consideration of a Bell scenario with two distinctive features: every party can measure two binary observables and the Bell inequality is only composed of full-correlators. These two features can be exploited—as in [42]—to drastically simplify the state shared by the parties without loss of generality and in a DI fashion. While the first feature allows the reduction to qubits, the second enables further simplifications on the multi-qubit state shared by the parties (for a reference, see Sect. 7.5).

Here we would like to provide an argument suggesting that any multipartite full-correlator Bell inequality with two binary measurements per party—e.g.. the MABK inequality—seems to be incompatible with the task of DICKA. We stress the fact that this is still an open question in the scientific community and there is not yet a formal proof which confirms or disproves the above statement. More details on this argument can be found in [42].

The secret conference key rate yielded by a generic N-partite DICKA protocol performed by Alice$_1$, ..., Alice$_N$, in the asymptotic limit, reads [10, 50]:

$$r_{\mathrm{DICKA}} = H(R_{A_1}|E) - \max_{2 \leq i \leq N} H(R_{A_1}|R_{A_i}). \tag{7.68}$$

The second term in (7.68) is due to EC (see Sect. 4.1) and represents the fact that Alice$_i$ for $i = 2, \ldots, N$ corrects her raw key to match Alice$_1$'s raw key. The conditional entropy $H(R_{A_1}|E)$ quantifies Eve's uncertainty on Alice$_1$'s key bits, which compose the secret conference key shared by all the parties after error correction and privacy amplification. As we discussed in the previous Section, the entropy $H(R_{A_1}|E)$ can be bounded when the parties observe a violation of an N-partite Bell inequality.

In light of the key rate expression (7.68), a DICKA protocol is successful (it can yield a positive key rate) when the following two events take place. The test-round data leads to a significant violation of a multiparty Bell inequality ($H(R_{A_1}|E)$ is large) and the parties' raw keys are sufficiently correlated ($H(R_{A_1}|R_{A_i})$ are small).

In the DIQKD protocol based on the CHSH inequality and illustrated in Sect. 7.4.2, one of the two test inputs of Alice is also used for key generation (KG), while Bob has a third additional input only devoted to KG. This fact is necessary in any DIQKD or DICKA protocol [10, 25]. In a DICKA protocol, we consider that Alice$_1$ plays the role of Alice, i.e. she is the only party without an input (observable) exclusively dedicated to KG.

If even Alice$_1$ had an additional setting only for KG, Eve—who manufactures the devices—would be able to distinguish a test round from a KG round on all the devices. Then, she could equip the devices with a maximally entangled state and suitable test-round measurements so that the parties would observe a maximal violation of the Bell inequality under test. Additionally, Eve could preprogram the devices to always output the same bit when the parties use their KG inputs, so that they would also have perfectly correlated raw keys. In doing so, Eve would be able to learn the whole conference key without being noticed.

The above argument implies that in an honest implementation of a DICKA protocol, the distributed quantum state and the chosen Bell inequality are such that the parties can have highly correlated outputs *while* violating the Bell inequality. Ideally, in an error-free implementation, it should be possible to maximally violate the Bell inequality and at the same time observe perfect correlations of the parties' raw keys. Indeed, this would maximize the protocol's asymptotic secret key rate (7.68) to: $r_{\text{DICKA}} = 1$.

Let us now consider a DICKA protocol based on the violation of a full-correlator Bell inequality with two binary observables per party. Here we heuristically argue that, for such DICKA protocols, it is forbidden to simultaneously have maximal Bell violation in the test rounds and perfect correlations in the KG rounds.

Given that the Bell inequality under consideration has two binary observables per party, we can restrict the analysis to multi-qubit states (c.f. Sect. 7.5). Then, we recall from Sect. 4.1 that the only multi-qubit state leading to perfectly correlated outcomes is the GHZ state (4.1) where every party measures the Pauli operator Z. If a party instead measures the operator X or Y, she would obtain a completely uncorrelated outcome. Therefore, we assume that the parties share a GHZ state and in the KG rounds every party measures the observable Z. In particular, this fixes one of Alice_1's test-round observables to be Z.

In [44], the authors show that every full-correlator Bell inequality with two binary observables per party is maximally violated by the GHZ state. However, we argue that in order to achieve maximal violation, the measurements must be chosen such that the resulting inequality (after simplifications) is only composed of expectation values of GHZ stabilizers (e.g., Eq. 7.22). Indeed, they acquire the extremal value 1 when evaluated on the GHZ state. Moreover, the stabilizers appearing in the inequality cannot contain the identity operator since that would not generate maximal violation. We identify these stabilizers as "full-stabilizers".

Unfortunately, all the observables of every N-partite GHZ state full-stabilizer (with N odd) are either the X or Y Pauli operators [24]. Therefore, in order to maximally violate the inequality, Alice_1's test-round observables lie in the (x, y)-plane of the Bloch sphere and have null Z component. This requirement collides with the fact that one of Alice_1's two observables is fixed to Z in order to have perfect correlations in KG rounds. A similar argument can be made for the N even case.[9]

Apparently, perfect correlations and maximal Bell violation are mutually exclusive conditions in any DICKA protocol based on a full-correlator Bell inequality with two binary observables per party, even for an ideal implementation of the protocol.

We emphasize that this argument, even if proven to be true, does not rule out the existence of implementations where the parties observe an adequate Bell violation while having reasonably correlated raw keys. However, it is an open question whether such implementations exist and lead to non-zero conference key rates.

---

[9]The $N = 2$ case includes the CHSH inequality that *can* be violated with Alice measuring Z, as we have seen in Sect. 7.4.2. However this is a degenerate case due to the low number of parties, as discussed in [42].

We point out that in [25] the authors have already discussed the apparent incompatibility of the tripartite MABK inequality with the task of DICKA. Indeed, they show that there exists no implementation such that the parties' outcomes are perfectly correlated and concurrently the MABK inequality is violated above the GME threshold, which is a necessary condition to ensure the privacy of the established key (c.f. Sect. 7.6.1).

In conclusion, the conditional entropy bounds presented in Sect. 7.6 are not likely to find direct application in the security of DICKA protocols. Nevertheless, since in DIRG the requirement of perfect correlations is dropped, they can still be employed in security proofs of DIRG protocols. Moreover, the techniques that led to the entropy bounds can inspire similar derivations which are relevant for the Bell inequalities used in current DICKA protocols [9, 10].

7.7.2 A Bell Inequality Tailored to DICKA

We conclude the Chapter by presenting a multipartite Bell inequality specifically designed to achieve both perfect correlations and maximal violation in an error-free implementation of a DICKA protocol. The Bell inequality is characterized by two binary observables per party, like in all the other cases discussed in this book. For the argument of the previous Subsection, the inequality is not exclusively composed of full-correlators.

The inequality under consideration has been introduced in [10] for the general case of N parties and its structure allows it to be maximally violated by an N-partite GHZ state where one of Alice$_1$'s optimal observables is Z. In this way, Alice$_1$'s outcomes are perfectly correlated with the other parties' outcomes, when every party measures Z in the KG rounds and the parties share a GHZ state.[10]

Here we focus on the $N = 3$ case for simplicity, where we denote the parties as Alice$_1$, Alice$_2$ and Alice$_3$ with observables $A_{x_i}^{(i)}$ for $i = 1, 2, 3$ and $x_i = 0, 1$. The Bell inequality in this case reads:

$$\langle A_1^{(1)} A_+^{(2)} A_+^{(3)} \rangle - \langle A_0^{(1)} A_-^{(2)} \rangle - \langle A_0^{(1)} A_-^{(3)} \rangle - \langle A_-^{(2)} A_-^{(3)} \rangle \leq 1, \tag{7.69}$$

where we defined non-normalized observables $A_{\pm}^{(j)} = (A_0^{(j)} \pm A_1^{(j)})/2$ for $j = 2, 3$. The maximal quantum violation is given by $3/2$ and is achieved on the tripartite GHZ state (2.30). The density operator relative to the tripartite GHZ state $|\text{GHZ}_3\rangle$ can be expressed in terms of all its stabilizers, similarly to (7.16), as follows [24]:

$$|\text{GHZ}_3\rangle\langle\text{GHZ}_3| = \frac{1}{8} (\mathbb{1} \otimes \mathbb{1} \otimes \mathbb{1} + Z \otimes Z \otimes \mathbb{1} + Z \otimes \mathbb{1} \otimes Z + \mathbb{1} \otimes Z \otimes Z$$
$$+ X \otimes X \otimes X - X \otimes Y \otimes Y - Y \otimes X \otimes Y - Y \otimes Y \otimes X). \tag{7.70}$$

---

[10]We recall that all the parties except for Alice$_1$ are equipped with a third measurement setting only used for KG.

The observables of Alice$_2$ and Alice$_3$, $A_x^{(j)} = \boldsymbol{\alpha}_x^{(j)} \cdot \boldsymbol{\sigma}$ (where $j = 2, 3$ and $\boldsymbol{\sigma} = (X, Y, Z)$), are qubit projective measurements[11] whose directions in the Bloch sphere are identified by the unit vectors $\boldsymbol{\alpha}_x^{(j)}$. Then, the non-normalized observables $A_+^{(j)}$ and $A_-^{(j)}$ read:

$$A_{\pm}^{(j)} = \boldsymbol{\alpha}_{\pm}^{(j)} \cdot \boldsymbol{\sigma} \quad , \quad \boldsymbol{\alpha}_{\pm}^{(j)} := \frac{\boldsymbol{\alpha}_0^{(j)} \pm \boldsymbol{\alpha}_1^{(j)}}{2} \tag{7.71}$$

and are characterized by orthogonal measurement directions $\boldsymbol{\alpha}_+^{(j)} \perp \boldsymbol{\alpha}_-^{(j)}$ and by normalizations which depend on each other: $\left|\boldsymbol{\alpha}_+^{(j)}\right|^2 + \left|\boldsymbol{\alpha}_-^{(j)}\right|^2 = 1$. These constraints must be taken into account when looking for the optimal observables leading to a maximal violation of (7.69).

Starting from the form of the inequality in (7.69), we can easily guess the optimal measurements to be performed on the GHZ state. In doing so, we follow the principle (see Eq. 7.22) that the resulting inequality should be only composed of correlators of the GHZ stabilizers (7.70). Note that the terms in (7.69) which are not full-correlators allow us to use the GHZ stabilizers containing the Z and the identity operator, without introducing the identity as one of the parties' observables. In this way we can impose that one of Alice$_1$'s optimal observables is Z, which is necessary to achieve perfect correlations with the other parties in the KG rounds.

For the arguments above, we choose the following optimal observables:

$$A_0^{(1)} = Z \quad , \quad A_1^{(1)} = X$$
$$A_-^{(j)} = -\frac{1}{2}Z \quad , \quad A_+^{(j)} = \frac{\sqrt{3}}{2}X \quad (j = 2, 3), \tag{7.72}$$

where $A_+^{(j)}$ and $A_-^{(j)}$ have orthogonal directions and Alice$_1$'s optimal observable $A_0^{(1)}$ is the Z operator. By substituting the optimal observables in (7.69) we obtain the maximal quantum violation:

$$\frac{3}{4}\langle XXX \rangle + \frac{1}{2}\langle ZZ \rangle + \frac{1}{2}\langle ZZ \rangle - \frac{1}{4}\langle ZZ \rangle = \frac{3}{2} > 1 \tag{7.73}$$

where all the terms of the inequality are indeed proportional to correlators of the GHZ stabilizers, which yield the value 1 when evaluated on the GHZ state.

The authors in [10] investigate the performance of the DICKA protocol based on the illustrated inequality. The security of the protocol is proven by bounding the single-round von Neumann entropy $H(R_A|E)$ as a function of the violation with rather loose numerical techniques [4, 39]. This inevitably leads to a poor performance of the conference key rate. A solution to this problem would be to derive tighter analytical bound on $H(R_A|E)$ similarly to what is done in [42] and possibly using

---

[11] We can restrict to qubit projective measurements since the Bell inequality has two inputs with binary outputs for each party.

similar techniques. The bound would guarantee a more accurate security analysis and hence a better performance.

Finally, we mention that the only other DICKA protocol proposed so far [9] is based on a Bell inequality that can be seen as a particular case of the one introduced in [10] and analysed here in the tripartite case. In particular, the inequality used in [9] is recovered when one imposes that Alice$_i$ for $i \geq 3$ has only one measurement setting at her disposal used for testing, instead of two.

The DICKA protocol proposed in [9] also lacks a tight security proof, and would benefit from tighter analytical bounds on $H(R_A|E)$ like the one derived in [42].

Appendix

In this Appendix we prove some statements made in the main text, whose articulated proof would have altered the cohesion and flow of the text.

7.8 State Reduction in the CHSH Scenario

Here we present the proof of Theorem 7.1, whose statement is reported for clarity.

Theorem 7.5 *([3]) Let Alice and Bob perform the DIQKD protocol described in Subsect. 7.4.2. It is not restrictive to assume that, in each round, Eve distributes a mixture $\sum_\alpha p_\alpha \rho_\alpha$ of two-qubit states ρ_α, together with a flag $|\alpha\rangle$ (known to her) which determines the measurements performed on ρ_α given the parties' inputs. Without loss of generality, the measurements performed by Alice's and Bob's devices on ρ_α are rank-one binary projective measurements in the (x, y)-plane of the Bloch sphere. Moreover, each state ρ_α is diagonal in the Bell basis (3.15) and reads:*

$$\rho_\alpha = \sum_{i,j=0}^{1} \lambda_{ij}^\alpha |\psi_{ij}\rangle\langle\psi_{ij}| \quad with \quad \lambda_{0j}^\alpha \geq \lambda_{1j}^\alpha \quad \forall j \in \{0, 1\}. \tag{7.74}$$

Proof The proof follows the same principles of the original proof in [3], however we apply some modifications and add details in a way which is coherent with its generalization valid for N parties that we prove in [42].

Reduction to qubits Firstly we reduce the state shared by Alice and Bob in one round to a convex combination of two-qubit states.

Recall that the statistics of a general quantum measurement (POVM, c.f. Sect. 2.2) is reproduced by a projective measurement in a larger Hilbert space, due to the Naimark theorem [55, 56]. Since in a DI scenario the Hilbert space dimensions are not fixed, we can assume without loss of generality (w.l.o.g.) that Alice and Bob perform binary projective measurements on their share of the quantum state.

Now we make use of a preliminary result derived in [3], which we extend and detail in [42]. The result states that the Hilbert space on which Alice's two projective measurements, corresponding to inputs $x = 0, 1$, are acting can be decomposed into the following direct sum (indicated by $\oplus$) of Hilbert spaces:

$$\mathcal{H} = \oplus_\alpha \mathcal{H}_\alpha^2 \, , \tag{7.75}$$

where every subspace $\mathcal{H}_\alpha^2$ is two-dimensional (qubit space) and both Alice's measurements act within $\mathcal{H}_\alpha^2$ as rank-one projective measurements.[12] Therefore, from Alice's perspective, the measurement process in one round consists of a projection in one of the two-dimensional subspaces $\mathcal{H}_\alpha^2$, followed by a projective measurement in that subspace selected according to her input. For this reason, we can think that Eve is effectively distributing to Alice a direct sum of qubits at every round. Moreover, since Eve fabricates the measurement devices, she can preprogram the projective measurements that Alice can select on every qubit. By repeating the same argument for Bob, we deduce that Eve effectively distributes a direct sum of two-qubit states in each round.

Now, consider that it cannot be detrimental for Eve to learn the value α corresponding to the two-qubit space selected in a particular round by Alice's and Bob's measurements. Hence, we can assume that Eve directly sends to the parties the two-qubit state ρ_α relative to the two-qubit space the parties would select. Since the selection of the two-qubit space can be random, Eve sends a statistical mixture of states ρ_α. Furthermore, since she could have preprogrammed the devices to perform specific measurements upon selecting a given subspace, together with the state ρ_α she sends a flag $|\alpha\rangle$ to Alice and Bob's devices to instruct them on which measurement to perform on the state ρ_α. In conclusion, in every round Eve prepares the following mixture of two-qubit states ρ_α:

$$\rho_{AB\Xi} = \sum_\alpha p_\alpha \rho_\alpha \otimes |\alpha\rangle\langle\alpha|_{\xi_A} \otimes |\alpha\rangle\langle\alpha|_{\xi_B}, \tag{7.76}$$

where the two ancillae $\Xi := \{\xi_A, \xi_B\}$ fix the qubit measurements that Alice and Bob can select on ρ_α. This can be modelled for instance by defining Alice's qubit measurement A_x as follows (and similarly Bob's):

$$A_x = \sum_\alpha \left(\Pi_{+1}^{x,\alpha} - \Pi_{-1}^{x,\alpha} \right) \otimes |\alpha\rangle\langle\alpha|_{\xi_A}, \tag{7.77}$$

---

[12] Note that a binary projective measurement on a qubit can also be of rank-two and corresponds to the identity as observable. In this case one outcome has probability 1 to occur and the other outcome never occurs. This possibility was originally neglected in [3] since measuring the identity cannot lead to a CHSH violation, as pointed out by [43]. However, the identity might lead to violations of multipartite Bell inequalities such as the MABK inequality we consider in [42]. In [42] we show how one can restrict to rank-one projective measurements, thus excluding the identity, without loss of generality.

where $\Pi_{\pm1}^{x,\alpha}$ are the projectors on the eigenvalues ±1 of the qubit projective measurement defined by Alice's choice of input x and by the particular state ρ_α she is measuring.

We now fix α and proceed in specifying the form of the two-qubit state ρ_α (we omit the symbol α in the following). At the moment, we can only say that ρ_α is a normalized positive Hermitian operator acting on the four-dimensional Hilbert space $\mathcal{H}_A \otimes \mathcal{H}_B$.

Symmetrization of the marginal distributions We define the planes individuated by the two measurements of Alice and of Bob to be the (x, y)-plane of the Bloch sphere. We can now assume w.l.o.g. that the marginal distributions of the outcomes, $p(a|x)$ and $p(b|y)$, are symmetrized. In other words, the expectation values of the corresponding observables are null:

$$\langle A_x \rangle = \langle B_y \rangle = 0 \quad \forall x, y \in \{0, 1\}, \tag{7.78}$$

where A_x and B_y are the observables of Alice and Bob, respectively, defining rank-one binary projective measurements in the (x, y)-plane. If (7.78) were not true, Alice and Bob could enforce it by agreeing on flipping their outcomes with probability $1/2$ in every round. This classical procedure would not change the observed CHSH value (7.27) nor the QBER, since either both parties flip or none of them does. Additionally, it would require classical communication between Alice and Bob, known to Eve.

Given that the experiment statistics satisfies (7.78), we can assume that it is Eve herself who flips the outcomes in place of the parties. However, instead of doing this classically on the outcomes of the devices, she could provide Alice and Bob with a suitable state which already embodies the symmetry of the outcomes distributions. By calling ρ the generic state she initially prepares, the state she distributes that satisfies (7.78) is given by:

$$\bar{\rho} = \frac{1}{2}\left[\rho + Z_A \otimes Z_B \, \rho \, Z_A^\dagger \otimes Z_B^\dagger\right], \tag{7.79}$$

where Z_A, Z_B are Pauli operators on Alice's and Bob's qubits. As a matter of fact, the outcome of a measurement in the (x, y)-plane is flipped if one first applies the Z operator.

We remark that it is safe to assume that Eve distributes the state (7.79) since this is not disadvantageous to her. Indeed, her uncertainty on Alice's raw key bit R_A, quantified by the conditional von Neumann entropy $H(R_A|E)$, does not increase when she sends the state $\bar{\rho}$ instead of ρ. The proof of this fact follows the same lines of the proof given in Sect. 3.5, hence we omit it.

By expressing the initial generic state ρ in the Bell basis (3.15):

$$\rho = \sum_{i,j,k,l=0}^{1} \rho_{(ij),(kl)} |\psi_{ij}\rangle\langle\psi_{kl}| \quad \rho_{(ij),(kl)} \in \mathbb{C}, \tag{7.80}$$

and by substituting it into (7.79), we observe that all the coherences relative to Bell states such that $j \neq l$ are set to zero:

$$\bar{\rho} = \sum_{i,j,k=0}^{1} \rho_{(ij),(kj)} |\psi_{ij}\rangle\langle\psi_{kj}|. \tag{7.81}$$

The matrix representation of the state in (7.81) in the Bell basis is thus block-diagonal and reads as follows, upon relabelling the coefficients[13]:

$$\bar{\rho} = \begin{bmatrix} \lambda_{00} & r_0 + is_0 & 0 & 0 \\ r_0 - is_0 & \lambda_{10} & 0 & 0 \\ 0 & 0 & \lambda_{01} & r_1 + is_1 \\ 0 & 0 & r_1 - is_1 & \lambda_{11} \end{bmatrix}, \tag{7.82}$$

where λ_{ij}, r_j and s_j are real numbers.

Exploiting the orientation of the local reference frames The state in (7.82) can be further reduced by carefully choosing the orientation of the parties' local reference frames. Indeed, although we already fixed the measurement directions of Alice and Bob to lie in the (x, y)-plane, we can still choose the orientation of the axes with respect to the measurement directions by applying rotations $R(\theta)$ along the z direction on the qubit spaces. In particular, the state distributed by Eve can be rotated w.l.o.g. as follows:

$$\bar{\rho}_+ = R_A(\theta_A) \otimes R_B(\theta_B) \, \bar{\rho} \, R_A^\dagger(\theta_A) \otimes R_B^\dagger(\theta_B), \tag{7.83}$$

where the rotation $R_A(\theta_A)$ acts on Alice's Hilbert space and is given by:

$$R_A(\theta_A) = \cos\left(\frac{\theta_A}{2}\right) \mathbb{1}_A + i \sin\left(\frac{\theta_A}{2}\right) Z_A, \tag{7.84}$$

and similarly for Bob. The resulting rotated state $\bar{\rho}_+$ is still block-diagonal and reads:

$$\bar{\rho}_+ = \begin{bmatrix} \lambda'_{00} & r_0 + is'_0 & 0 & 0 \\ r_0 - is'_0 & \lambda'_{10} & 0 & 0 \\ 0 & 0 & \lambda'_{01} & r_1 + is'_1 \\ 0 & 0 & r_1 - is'_1 & \lambda'_{11} \end{bmatrix}, \tag{7.85}$$

where the new matrix coefficients are given by:

---

[13] Recall that $\bar{\rho}$ is a Hermitian operator, hence the matrix representing it must be Hermitian.

$$\lambda'_{ij} = \frac{1}{2} \left[\lambda_{0j} + \lambda_{1j} + (-1)^i (\lambda_{0j} - \lambda_{1j}) \cos[\theta_j(\theta_A, \theta_B)] + 2(-1)^i s_j \sin[\theta_j(\theta_A, \theta_B)] \right]$$
$$(7.86)$$

$$s'_j = s_j \cos[\theta_j(\theta_A, \theta_B)] - \frac{1}{2}(\lambda_{0j} - \lambda_{1j}) \sin[\theta_j(\theta_A, \theta_B)], \tag{7.87}$$

where the angle $\theta_j(\theta_A, \theta_B)$ is defined as:

$$\theta_j(\theta_A, \theta_B) := \theta_A + (-1)^j \theta_B. \tag{7.88}$$

From (7.87) we deduce that by choosing the rotation angles such that both the following conditions are verified[14]:

$$\theta_A + (-1)^j \theta_B = \arctan \frac{2s_j}{\lambda_{0j} - \lambda_{1j}} \quad \text{for } j = 0, 1, \tag{7.89}$$

we can set the imaginary parts of the off-diagonal terms to zero: $s'_0 = s'_1 = 0$. Thus, w.l.o.g. we can assume that the state distributed by Eve is of the form:

$$\bar{\rho}_+ = \begin{bmatrix} \lambda_{00} & r_0 & 0 & 0 \\ r_0 & \lambda_{10} & 0 & 0 \\ 0 & 0 & \lambda_{01} & r_1 \\ 0 & 0 & r_1 & \lambda_{11} \end{bmatrix}. \tag{7.90}$$

Moreover, by applying further rotations on (7.90) defined by angles $\tilde{\theta}_A$ and $\tilde{\theta}_B$ such that:

$$\tilde{\theta}_A + (-1)^j \tilde{\theta}_B = \pi, \tag{7.91}$$

we can exchange the position of the two diagonal terms λ_{0j} and λ_{1j} in (7.90), for $j = 0, 1$ (see (7.86)). This implies that we can assume w.l.o.g. that the diagonal elements in (7.90) are ordered as follows:

$$\lambda_{00} \geq \lambda_{10} \quad, \quad \lambda_{01} \geq \lambda_{11}. \tag{7.92}$$

Independence from the off-diagonal terms Finally, let us construct the state $\bar{\rho}_-$ starting from $\bar{\rho}_+$ given in (7.90) by replacing r_j with $-r_j$:

$$\bar{\rho}_- := \begin{bmatrix} \lambda_{00} & -r_0 & 0 & 0 \\ -r_0 & \lambda_{10} & 0 & 0 \\ 0 & 0 & \lambda_{01} & -r_1 \\ 0 & 0 & -r_1 & \lambda_{11} \end{bmatrix}. \tag{7.93}$$

---

[14]Note that this is possible since we have two linear conditions for two independent variables.

We observe that the two states $\bar{\rho}_+$ and $\bar{\rho}_-$ yield the same probability distribution of the outcomes:

$$p(a,b)_{\bar{\rho}_+} = \mathrm{Tr}\left[\Pi_a \Pi_b \bar{\rho}_+\right] = \mathrm{Tr}\left[\Pi_a \Pi_b \bar{\rho}_-\right] = p(a,b)_{\bar{\rho}_-}, \qquad (7.94)$$

where the projectors Π_a and Π_b represent Alice and Bob's projective measurements in the (x, y)-plane relative to some non-specified inputs. The projectors can be parametrized by writing the corresponding observables A and B as convex combinations of the Pauli operators X and Y:

$$\begin{aligned} A &= \cos(\varphi_A)X + \sin(\varphi_A)Y \\ B &= \cos(\varphi_B)X + \sin(\varphi_B)Y, \end{aligned} \qquad (7.95)$$

for some unknown angles φ_A, φ_B. The eigenstates of the observables in (7.95) read:

$$\begin{aligned} |a\rangle_A &= \frac{1}{\sqrt{2}}(|0\rangle + (-1)^a e^{i\varphi_A}|1\rangle) \\ |b\rangle_B &= \frac{1}{\sqrt{2}}(|0\rangle + (-1)^b e^{i\varphi_B}|1\rangle) \end{aligned} \qquad (7.96)$$

where the measurement outcomes are defined as $a, b \in \{0, 1\}$ ($a = 0$ corresponds to eigenvalue $+1$ and $a = 1$ to eigenvalue -1). Then the projectors Π_a and Π_b are simply given by $\Pi_a = |a\rangle\langle a|_A$ and $\Pi_b = |b\rangle\langle b|_B$.

Furthermore, the states $\bar{\rho}_+$ and $\bar{\rho}_-$ provide Eve with the same information, i.e. their conditional entropies coincide:

$$H(R_A|E)_{\bar{\rho}_+} = H(R_A|E)_{\bar{\rho}_-}. \qquad (7.97)$$

Additionally, it is not disadvantageous for Eve to prepare the balanced mixture:

$$\rho_\alpha := \frac{\bar{\rho}_+ + \bar{\rho}_-}{2}, \qquad (7.98)$$

rather than preparing one of the two states with certainty, if she knows which of the two states she prepared:

$$H(R_A|E)_{\rho_\alpha} \leq H(R_A|E)_{\bar{\rho}_+}. \qquad (7.99)$$

The proofs of the observations (7.94), (7.97) and (7.99) follow by direct computation and are omitted. Nevertheless, the interested reader can find analogous proofs in the Supplementary Information in [42], valid for the general N-party scenario.

We conclude that it is not restrictive to assume that Eve distributes to the parties the mixture (7.76) of two-qubit states ρ_α together with ancillae that fix the parties' possible measurements. Each state ρ_α is given by (7.98) and is diagonal in the Bell basis:

$$\rho_\alpha = \begin{bmatrix} \lambda_{00} & 0 & 0 & 0 \\ 0 & \lambda_{10} & 0 & 0 \\ 0 & 0 & \lambda_{01} & 0 \\ 0 & 0 & 0 & \lambda_{11} \end{bmatrix}, \tag{7.100}$$

with the conditions (7.92) on the diagonal elements. This concludes the proof.

7.9 Lower Bound on the Conditional Entropy: Analytical Proof

The security of the DIQKD protocol presented in Sect. 7.4.2 is based on the ability to lower bound the conditional von Neumann entropy $H(R_A|E)_{\rho_\alpha}$ as a function of the CHSH violation S_α, as discussed in Sect. 7.5. There, the conditional entropy is simplified to (7.49):

$$H(R_A|E)_{\rho_\alpha} = 1 - H(\{\lambda_{ij}^\alpha\}) + h(\lambda_{00}^\alpha + \lambda_{01}^\alpha), \tag{7.101}$$

and its lower bound is obtained by solving the following optimization problem:

$$F(S_\alpha) := \min_{\{\lambda_{ij}^\alpha\}} 1 - H(\{\lambda_{ij}^\alpha\}) + h(\lambda_{00}^\alpha + \lambda_{01}^\alpha)$$

$$\text{sub. to} \quad S_\alpha \geq \bar{S}_\alpha \; ; \; \lambda_{0j}^\alpha \geq \lambda_{1j}^\alpha \; ; \; \sum_{i,j=0,1} \lambda_{ij}^\alpha = 1 \tag{7.102}$$

where $\bar{S}_\alpha$ is the maximal CHSH violation given in (7.33) and reported here for completeness:

$$\bar{S}_\alpha = 2\sqrt{2} \max \left\{ \sqrt{(\lambda_{00}^\alpha - \lambda_{11}^\alpha)^2 + (\lambda_{01}^\alpha - \lambda_{10}^\alpha)^2}, \sqrt{(\lambda_{00}^\alpha - \lambda_{10}^\alpha)^2 + (\lambda_{01}^\alpha - \lambda_{11}^\alpha)^2} \right\}. \tag{7.103}$$

Here we analytically derive the solution of the optimization problem in (7.102). We start by assuming that the CHSH value S_α is such that $S_\alpha \geq 2$, i.e. we assume that the CHSH inequality is violated. Otherwise, Eve would have full information on Alice's raw key bit R_A and the lower bound on the conditional entropy would be zero.

Because of the symmetry of the problem, we assume w.l.o.g. that $\lambda_{01}^\alpha \geq \lambda_{00}^\alpha$. Indeed, for every solution of (7.102) with $\lambda_{00}^\alpha \geq \lambda_{01}^\alpha$, there exists an equivalent solution—that leads to the same minimum—with $\lambda_{01}^\alpha \geq \lambda_{00}^\alpha$: the equivalent solution is obtained by relabelling $\lambda_{01}^\alpha \leftrightarrow \lambda_{00}^\alpha$.[15]

---

[15] Note that the relabelling does not modify the maximal CHSH violation (7.103).

By noticing that the second term in (7.103) is larger than the first if and only if $\lambda_{01}^\alpha \geq \lambda_{00}^\alpha$, we can simplify the maximal CHSH violation to:

$$S_\alpha = 2\sqrt{2}\sqrt{(\lambda_{00}^\alpha - \lambda_{10}^\alpha)^2 + (\lambda_{01}^\alpha - \lambda_{11}^\alpha)^2}. \qquad (7.104)$$

Then, a necessary condition for $S_\alpha \geq 2$ can be derived by upper bounding (7.104) as follows:

$$2 \leq S_\alpha \leq S_\alpha \leq 2\sqrt{2}\sqrt{(\lambda_{00}^\alpha)^2 + (\lambda_{01}^\alpha)^2} \leq 2\sqrt{2}\sqrt{\lambda_{01}^\alpha(\lambda_{01}^\alpha + \lambda_{00}^\alpha)} \leq 2\sqrt{2}\sqrt{\lambda_{01}^\alpha}, \qquad (7.105)$$

which implies the following necessary condition on λ_{01}^α:

$$\lambda_{01}^\alpha \geq \frac{1}{2}. \qquad (7.106)$$

Consider the following class of states parametrized by $\nu \in [\frac{1}{2}, 1]$:

$$\tau(\nu) = (1 - \nu)|\psi_{00}\rangle\langle\psi_{00}| + \nu|\psi_{01}\rangle\langle\psi_{01}|, \qquad (7.107)$$

whose maximal CHSH violation (7.103) reads:

$$S_\tau(\nu) = 2\sqrt{2}\sqrt{\nu^2 + (1 - \nu)^2}. \qquad (7.108)$$

It is straightforward to verify, by using the last expression, that

$$S_\tau(\lambda_{01}^\alpha) \geq S_\alpha \quad \forall \rho_\alpha, \qquad (7.109)$$

where S_α is given in (7.104). Moreover, the entropy (7.101) of the states (7.107) reads:

$$H(X|E)_\tau(\nu) = 1 - h(\nu), \qquad (7.110)$$

where we used the binary entropy $h(x) = -x \log x - (1 - x)\log(1 - x)$.

By definition of the optimization problem (7.102), the solution of the optimization for a given S_α is upper bounded by the entropy of any particular state with $S_\alpha = S_\alpha$. Thus for the states (7.107) we have:

$$F(S_\alpha) \leq H(R_A|E)_\tau(\nu_\alpha) \qquad (7.111)$$

where ν_α is fixed such that the maximal violation $S_\tau(\nu_\alpha)$ of the state $\tau(\nu_\alpha)$ is exactly given by S_α:

$$S_\tau(\nu_\alpha) = 2\sqrt{2}\sqrt{\nu_\alpha^2 + (1 - \nu_\alpha)^2} = S_\alpha, \qquad (7.112)$$

where we choose the solution $\nu_\alpha \geq 1/2$.

By proving the following result (with the assumption $\lambda_{01}^{\alpha} \geq \lambda_{00}^{\alpha}$ and (7.106)):

$$H(R_A|E)_{\rho_{\alpha}} \geq H(R_A|E)_{\tau}(\lambda_{01}^{\alpha}) \quad \forall \rho_{\alpha} \tag{7.113}$$

we obtain the solution of the optimization problem. In order to realize this, let us use the last expression on the state ρ_{α}^{*}, which is the solution of the minimization in (7.102):

$$F(S_{\alpha}) = H(R_A|E)_{\rho_{\alpha}^{*}} \geq H(R_A|E)_{\tau}(\lambda_{01}^{\alpha,*})$$
$$\geq H(R_A|E)_{\tau}(\nu_{\alpha}). \tag{7.114}$$

The last inequality in (7.114) is motivated by the following two observations:

- By applying (7.109) to the state ρ_{α}^{*}, we obtain $S_{\tau}(\lambda_{01}^{\alpha,*}) \geq S_{\alpha,*} \geq S_{\alpha}$, which combined with (7.112) implies that $\lambda_{01}^{\alpha,*} \geq \nu_{\alpha}$ since $S_{\tau}(\nu)$ in (7.108) is monotonically increasing in the interval $\nu \in [\frac{1}{2}, 1]$.
- The entropy $H(R_A|E)_{\tau}(\nu)$ in (7.110) is monotonically increasing in the interval $\nu \in [\frac{1}{2}, 1]$.

The above observations lead to the second inequality in (7.114).

By combining (7.114) with (7.111), we obtain the desired lower bound:

$$F(S_{\alpha}) = H(R_A|E)_{\tau}(\nu_{\alpha}) = 1 - h(\nu_{\alpha})$$
$$= 1 - h\left(\frac{1}{2} + \frac{1}{2}\sqrt{\left(\frac{S_{\alpha}}{2}\right)^2 - 1}\right), \tag{7.115}$$

where the last equality is obtained by reverting Eq. (7.112).

The bound (7.115) is a *tight* solution of the optimization problem in (7.102). Indeed, for every violation S_{α}, there exists a state $\tau(\nu_{\alpha})$ such that its entropy coincides with the bound and such that it can produce a violation equal to S_{α}, thanks to $S_{\tau}(\nu_{\alpha}) = S_{\alpha}$ (7.112) and to the fact that the maximal CHSH violation (7.103) is achievable.

We are left to prove the inequality in (7.113), which can be made explicit by using (7.101) and (7.110):

$$D := h(\lambda_{01}^{\alpha}) - H(\{\lambda_{ij}^{\alpha}\}) + h(\lambda_{00}^{\alpha} + \lambda_{01}^{\alpha}) \geq 0. \tag{7.116}$$

We simplify the first two terms in D:

$$h(\lambda_{01}^{\alpha}) - H(\{\lambda_{ij}^{\alpha}\}) = -(1 - \lambda_{01}^{\alpha})\log(1 - \lambda_{01}^{\alpha}) + \sum_{(i,j)\neq(0,1)} \lambda_{ij}^{\alpha} \log \lambda_{ij}^{\alpha}. \tag{7.117}$$

Now we apply Jensen's inequality:

$$f(x+y) \geq \frac{f(2x)+f(2y)}{2} \quad \text{for} \quad f(x) = -x\log x, \quad (7.118)$$

to the last term in (7.116):

$$\begin{aligned}
h(\lambda_{00}^\alpha + \lambda_{01}^\alpha) &= -(\lambda_{00}^\alpha + \lambda_{01}^\alpha)\log(\lambda_{00}^\alpha + \lambda_{01}^\alpha) + f(\lambda_{10}^\alpha + \lambda_{11}^\alpha) \\
&\geq -(\lambda_{00}^\alpha + \lambda_{01}^\alpha)\log(\lambda_{00}^\alpha + \lambda_{01}^\alpha) - \lambda_{10}^\alpha \log(2\lambda_{10}^\alpha) - \lambda_{11}^\alpha \log(2\lambda_{11}^\alpha) \\
&= -(\lambda_{00}^\alpha + \lambda_{01}^\alpha)\log(\lambda_{00}^\alpha + \lambda_{01}^\alpha) - (1 - \lambda_{00}^\alpha - \lambda_{01}^\alpha) \\
&\quad - \lambda_{10}^\alpha \log(\lambda_{10}^\alpha) - \lambda_{11}^\alpha \log(\lambda_{11}^\alpha). \quad (7.119)
\end{aligned}$$

By combining (7.117) and (7.119) in (7.116) we get:

$$\begin{aligned}
D &\geq -(1 - \lambda_{01}^\alpha)\log(1 - \lambda_{01}^\alpha) - (\lambda_{00}^\alpha + \lambda_{01}^\alpha)\log(\lambda_{00}^\alpha + \lambda_{01}^\alpha) - (1 - \lambda_{00}^\alpha - \lambda_{01}^\alpha) \\
&\quad + \lambda_{00}^\alpha \log \lambda_{00}^\alpha \\
&= -(\lambda_{00}^\alpha + \lambda_{01}^\alpha)\log(\lambda_{00}^\alpha + \lambda_{01}^\alpha) - (1 - \lambda_{01}^\alpha)\log[2(1 - \lambda_{01}^\alpha)] + \lambda_{00}^\alpha \log(2\lambda_{00}^\alpha) \\
&=: g(\lambda_{01}^\alpha, \lambda_{00}^\alpha). \quad (7.120)
\end{aligned}$$

In the last expression we defined the function $g(x, y)$:

$$g(x, y) = -(x + y)\log(x + y) - (1 - x)\log[2(1 - x)] + y\log(2y), \quad (7.121)$$

and we will analyse it in the ranges of interest for the variables $x = \lambda_{01}^\alpha$ and $y = \lambda_{00}^\alpha$, i.e.: $1/2 \leq x \leq 1, 0 \leq y \leq 1 - x$.

In these ranges the function in (7.121) is concave in x since its second derivative is always negative:

$$\frac{\partial^2 g(x, y)}{\partial x^2} = -\frac{1}{\ln(2)}\left(\frac{1}{1 - x} + \frac{1}{x + y}\right) < 0. \quad (7.122)$$

Consider the points at the boundary $x + y = 1$, for which we get $g(1 - y, y) = 0$. Thanks to the concavity of $g(x, y)$, it holds that:

$$g\left(p\frac{1}{2} + (1 - p)(1 - y), y\right) \geq pg\left(\frac{1}{2}, y\right) + (1 - p)g(1 - y, y), \quad 0 \leq p \leq 1$$

or equivalently that:

$$g(x, y) \geq \left(\frac{1 - x - y}{\frac{1}{2} - y}\right)g\left(\frac{1}{2}, y\right). \quad (7.123)$$

Note that in the parameter regimes of x and y it holds that

$$0 \le \left(\frac{1 - x - y}{\frac{1}{2} - y} \right) \le 1. \tag{7.124}$$

We finally analyse the properties of $g(\frac{1}{2}, y)$, which is convex in y since its second derivative is always positive:

$$\frac{\partial^2 g(\frac{1}{2}, y)}{\partial y^2} = \frac{1}{y \ln(2) + y^2 \ln(4)} > 0. \tag{7.125}$$

A convex function has a unique minimum if it exists in the parameter regime. In our case this is given by:

$$\frac{\partial g(\frac{1}{2}, y)}{\partial y} = \log(2y) - \log(\frac{1}{2} + y) \overset{!}{=} 0 \quad \Leftrightarrow \quad y = \frac{1}{2} \tag{7.126}$$

for which $g(\frac{1}{2}, \frac{1}{2}) = 0$ holds. Thus in general it holds:

$$g\left(\frac{1}{2}, y \right) \ge 0. \tag{7.127}$$

By combining these considerations we prove the inequality in (7.116):

$$D \overset{(7.120)}{\ge} g(\lambda_{01}^\alpha, \lambda_{00}^\alpha)$$

$$\overset{(7.123)}{\ge} \left(\frac{1 - \lambda_{01}^\alpha - \lambda_{00}^\alpha}{\frac{1}{2} - \lambda_{00}^\alpha} \right) g\left(\frac{1}{2}, \lambda_{00}^\alpha \right)$$

$$\ge 0, \tag{7.128}$$

where in the last inequality we used the fact that the pre-factor is positive (7.124) and that $g(\frac{1}{2}, \lambda_{00}^\alpha)$ is lower bounded by zero (7.127). This concludes the proof of inequality (7.113), thus completing the analytical solution of the optimization problem in (7.102).

References

1. Yao, A., Mayers, D., & (1998). Quantum cryptography with imperfect apparatus. In *2013 IEEE 54th Annual Symposium on Foundations of Computer Science, , Los Alamitos, CA* (p. 503). USA: IEEE Computer Society.
2. Acín, A., Gisin, N., & Masanes, L. (2006). From Bell's theorem to secure quantum key distribution. *Physical Review Letters, 97*, 120405.

3. Pironio, S., Acín, A., Brunner, N., Gisin, N., Massar, S., & Scarani, V. (2009). Device-independent quantum key distribution secure against collective attacks. *New Journal of Physics*, *11*(4), 045021.
4. Masanes, L., Pironio, S., & Acín, A. (2011). Secure device-independent quantum key distribution with causally independent measurement devices. *Nature Communications*, *2*(1), 238.
5. Vazirani, U., & Vidick, T. (2014). Fully device-independent quantum key distribution. *Physical Review Letters*, *113*, 140501.
6. Arnon-Friedman, R., Dupuis, F., Fawzi, O., Renner, R., & Vidick, T. (2018). Practical device-independent quantum cryptography via entropy accumulation. *Nature Communications*, *9*(1), 459.
7. Scarani, V., & Gisin, N. (2001a). Quantum communication between n partners and Bell's inequalities. *Physical Review Letters*, *87*, 117901.
8. Scarani, V., & Gisin, N. (2001b). Quantum key distribution between n partners: Optimal eavesdropping and Bell's inequalities. *Physical Review A*, *65*, 012311.
9. Ribeiro, J., Murta, G., & Wehner, S. (2019). Reply to "comment on 'fully device-independent conference key agreement'". *Physical Review A*, *100*, 026302.
10. Holz, T., Kampermann, H., & Bruß, D. (2019). A genuine multipartite bell inequality for device-independent conference key agreement. arXiv:quant-ph/1910.11360.
11. Colbeck, R. (2007). Quantum and relativistic protocols for secure multi-party computation. Ph.D. thesis, University of Cambridge. arXiv:quant-ph/0911.3814.
12. Pironio, S., Acín, A., Massar, S., de la Giroday, A. B., Matsukevich, D. N., Maunz, P., et al. (2010). Random numbers certified by Bell's theorem. *Nature*, *464*(7291), 1021–1024.
13. Colbeck, R., & Kent, A. (2011). Private randomness expansion with untrusted devices. *Journal of Physics A: Mathematical and Theoretical*, *44*(9), 095305.
14. Nieto-Silleras, O., Bamps, C., Silman, J., & Pironio, S. (2018). Device-independent randomness generation from several bell estimators. *New Journal of Physics*, *20*(2), 023049.
15. Pironio, S., & Massar, S. (2013). Security of practical private randomness generation. *Physical Review A*, *87*, 012336.
16. Fehr, S., Gelles, R., & Schaffner, C. (2013). Security and composability of randomness expansion from bell inequalities. *Physical Review A*, *87*, 012335.
17. Bell, J. S. (1964). On the Einstein podolsky Rosen paradox. *Physics Physique Fizika*, *1*, 195–200.
18. Bell, J. S. (2004). *Speakable and Unspeakable in Quantum Mechanics*. Cambridge: Cambridge University Press.
19. Valdenebro, A. G. (2002). Assumptions underlying Bell's inequalities. *European Journal of Physics*, *23*(5), 569–577.
20. Goldstein, S., Norsen, T., Tausk, D. V., & Zanghi, N. (2011). Bell's theorem. *Scholarpedia*, *6*(10), 8378. Revision #91049.
21. Brunner, N., Cavalcanti, D., Pironio, S., Scarani, V., & Wehner, S. (2014). Bell nonlocality. *Reviews of Modern Physics*, *86*, 419–478.
22. Horodecki, P., & Ramanathan, R. (2019). The relativistic causality versus no-signaling paradigm for multi-party correlations. *Nature Communications*, *10*(1), 1701.
23. Clauser, J. F., Horne, M. A., Shimony, A., & Holt, R. A. (1969). Proposed experiment to test local hidden-variable theories. *Physical Review Letters*, *23*, 880–884.
24. Tóth, G., & Gühne, O. (2005). Entanglement detection in the stabilizer formalism. *Physical Review A*, *72*, 022340.
25. Holz, T., Miller, D., Kampermann, H., & Bruß, D. (2019). Comment on "fully device-independent conference key agreement". *Physical Review A*, *100*, 026301.
26. Aspect, A., Dalibard, J., & Roger, G. (1982). Experimental test of Bell's inequalities using time-varying analyzers. *Physical Review Letters*, *49*, 1804–1807.
27. Hensen, B., Bernien, H., Dréau, A. E., Reiserer, A., Kalb, N., Blok, M. S., Ruitenberg, J., Vermeulen, R. F. L., Schouten, R. N., Abellán, C., Amaya, W., Pruneri, V., Mitchell, M. W., Markham, M., Twitchen, D. J., Elkouss, D., Wehner, S., Taminiau, T. H., & Hanson, R. (2015). Loophole-free bell inequality violation using electron spins separated by 1.3 kilometres. *Nature*, *526*(7575), 682–686.

28. Giustina, M., Versteegh, M. A. M., Wengerowsky, S., Handsteiner, J., Hochrainer, A., Phelan, K., et al. (2015). Significant-loophole-free test of Bell's theorem with entangled photons. *Physical Review Letters, 115*, 250401.
29. Shalm, L. K., Meyer-Scott, E., Christensen, B. G., Bierhorst, P., Wayne, M. A., Stevens, M. J., et al. (2015). Strong loophole-free test of local realism. *Physical Review Letters, 115*, 250402.
30. Masanes, L., Acin, A., & Gisin, N. (2006). General properties of nonsignaling theories. *Physical Review A, 73*, 012112.
31. Coffman, V., Kundu, J., & Wootters, W. K. (2000). Distributed entanglement. *Physical Review A, 61*, 052306.
32. Barrett, J., Kent, A., & Pironio, S. (2006). Maximally nonlocal and monogamous quantum correlations. *Physical Review Letters, 97*, 170409.
33. Dupuis, F., Fawzi, O., & Renner, R. (2016). Entropy accumulation. arXiv:quant-ph/1607.01796.
34. Dupuis, F., & Fawzi, O. (2019). Entropy accumulation with improved second-order term. *IEEE Transactions on Information Theory, 65*(11), 7596–7612.
35. Murta, G., van Dam, S. B., Ribeiro, J., Hanson, R., & Wehner, S. (2019). Towards a realization of device-independent quantum key distribution. *Quantum Science and Technology, 4*(3), 035011.
36. Pirandola, S., Andersen, U. L., Banchi, L., Berta, M., Bunandar, D., Colbeck, R., Englund, D., Gehring, T., Lupo, C., Ottaviani, C., Pereira, J., Razavi, M., Shaari, J. S., Tomamichel, M., Usenko, V. C., Vallone, G., Villoresi, P., & Wallden, P. (2019). Advances in quantum cryptography. arXiv:quant-ph/1906.01645.
37. Arnon-Friedman, R., Renner, R., & Vidick, T. (2019). Simple and tight device-independent security proofs. *SIAM Journal on Computing, 48*(1), 181–225.
38. Navascués, M., Pironio, S., & Acín, A. (2007). Bounding the set of quantum correlations. *Physical Review Letters, 98*, 010401.
39. Navascués, M., Pironio, S., & Acín, A. (2008). A convergent hierarchy of semidefinite programs characterizing the set of quantum correlations. *New Journal of Physics, 10*(7), 073013.
40. Nieto-Silleras, O., Pironio, S., & Silman, J. (2014). Using complete measurement statistics for optimal device-independent randomness evaluation. *New Journal of Physics, 16*(1), 013035.
41. Bancal, J.-D., Sheridan, L., & Scarani, V. (2014). More randomness from the same data. *New Journal of Physics, 16*(3), 033011.
42. Grasselli, F., Murta, G., Kampermann, H., & Bruß, D. (2020). Analytical entropic bounds for multiparty device-independent cryptography. arXiv:quant-ph/2004.14263.
43. Horodecki, R., Horodecki, P., & Horodecki, M. (1995). Violating bell inequality by mixed spin-12 states: Necessary and sufficient condition. *Physics Letters A, 200*(5), 340–344.
44. Werner, R. F., & Wolf, M. M. (2001). All-multipartite bell-correlation inequalities for two dichotomic observables per site. *Physical Review A, 64*, 032112.
45. Epping, M., Kampermann, H., Macchiavello, C., & Bruß, D. (2017). Multi-partite entanglement can speed up quantum key distribution in networks. *New Journal of Physics, 19*(9), 093012.
46. Mermin, N. D. (1990). Extreme quantum entanglement in a superposition of macroscopically distinct states. *Physical Review Letters, 65*, 1838–1840.
47. Ardehali, M. (1992). Bell inequalities with a magnitude of violation that grows exponentially with the number of particles. *Physical Review A, 46*, 5375–5378.
48. BelinskiĭÂ, A. V., & Klyshko, D. N. (1993). Interference of light and Bell's theorem. *Physical Review A, 36*, 653–693.
49. Collins, D., Gisin, N., Popescu, S., Roberts, D., & Scarani, V. (2002). Bell-type inequalities to detect true *n*-body nonseparability. *Physical Review Letters, 88*, 170405.
50. Ribeiro, J., Murta, G., & Wehner, S. (2018). Fully device-independent conference key agreement. *Physical Review A, 97*, 022307.
51. Siddiqui, M. A., & Sazim, S. (2019). Tight upper bound for the maximal expectation value of the mermin operators. *Quantum Information Processing, 18*(5), 131.
52. Brown, P. J., Ragy, S., & Colbeck, R. (2020). A framework for quantum-secure device-independent randomness expansion. *IEEE Transactions on Information Theory, 66*(5), 2964–2987.

53. Woodhead, E., Bourdoncle, B., & Acín, A. (2018). Randomness versus nonlocality in the Mermin-Bell experiment with three parties. *Quantum, 2,* 82.
54. Carrara, G., Kampermann, H., Bruß, D., & Murta, G. (2020). Genuine multipartite entanglement is not a precondition for secure conference key agreement. arXiv:quant-ph/2007.11553.
55. Peres, A. (2006). *Quantum Theory: Concepts and Methods.* Dordrecht: Springer.
56. Paris, M. G. A. (2012). The modern tools of quantum mechanics. *The European Physical Journal Special Topics, 203*(1), 61–86.

Chapter 8
Conclusion and Outlook

The fruitful combination of concepts and ideas from different fields of study, as well as the practical implications for the security of our data, have made quantum cryptography a very active research topic in recent years. In this context a major role is played by quantum key distribution (QKD) [1], which enables information-theoretic secure communication between two users. In a world evermore demanding for connectedness, the need for an equally-secure communication established among several users is going to be satisfied by quantum conference key agreement (CKA) [2].

CKA is only one aspect of a wider vision on how quantum communication will change our lives, with the quantum internet as its most ambitious representative [3, 4]. Within this view, future quantum networks will provide on-demand entanglement to any subset of users in the network, allowing the execution of quantum-enabled tasks unachievable with classical means. Prominent examples include: blind quantum computing (the distribution of quantum computations to remote quantum servers, without them knowing the nature of the computation) [5, 6], clocks synchronization [7], and quantum anonymous voting [8, 9].

The recent developments illustrated in this book have contributed to driving the transition of quantum-secured communication beyond the two-user paradigm, from bipartite QKD to CKA.

As a matter of fact, we presented the generalization of the QKD security definitions to the multipartite scenario, allowing the analysis of CKA schemes in the finite-key regime [10]. We discussed in detail two CKA protocols [10, 11], provided insight on their security proof and benchmarked their performance with other CKA schemes and with the iteration of bipartite QKD protocols.

Despite the book being mainly focused on theoretical aspects, we provided a brief overview of the state-of-the-art experiments on QKD and CKA. Moreover, we described in detail the first experimental implementation of a CKA [12] and highlighted the challenges in implementing other CKA schemes.

F. Grasselli, *Quantum Cryptography*, Quantum Science and Technology,
https://doi.org/10.1007/978-3-030-64360-7_8

At the same time, the book attempts to provide a rather comprehensive overview on the topic of QKD, from its origins to its most advanced protocols. New or improved QKD protocols are being developed at an astonishing pace and it is quite difficult to cover all of them in a satisfactory way.

One of the promising new protocols we focus on is TF-QKD, which has arguably become the new benchmark for far-distance QKD. We showed that TF-QKD is a major candidate for being implemented in near-future quantum networks. Additionally, we dedicate a Chapter to device-independent (DI) protocols, which guarantee the highest level of security in quantum cryptography. In this context we focus on the security aspects of the protocols and illustrate novel theoretical tools for the security of multipartite DI protocols [13].

CKA is arguably in its infancy and there is still much to be done in order to concretely make CKA protocols the ultimate solution for secure multi-user communication. Here we briefly outline some research directions that may be pursued.

As discussed in the book, multipartite entanglement seems to be a necessary ingredient of CKA protocols. However, the distribution of multipartite entangled states to the participating parties is not an easy task and it often requires the coincident arrival of each photon to the corresponding party. This fact limits the maximum distance between any pair of parties due to photon loss. Advances both in CKA design and distribution of multipartite entanglement can mitigate this issue and allow CKA to achieve longer distances (see e.g., Sect. 6.5).

From an experimental point of view, the described CKA experiments are only the first step towards a fully fledged CKA which can serve the needs of secure multi-user communication. We point out two aspects that should be addressed to meet such a goal. Firstly, one should increase the generation rate of the distributed multipartite entangled state in order to speed up the resulting secure communication. Additionally, the future field implementation of CKA should be performed in the existing telecommunication infrastructure. This would remove the need for dedicated fibre networks linking the users.

A theoretical aspect where CKA is still quite underdeveloped are DI protocols (DICKA protocols). The tools developed in [13] lay the ground for the derivation of entropy bounds for the security of multipartite DI cryptographic protocols. We remark that a careful estimation of these entropies is of paramount importance for the experimental feasibility of DI protocols, as it increases noise tolerance and relaxes the experimental requirements. Of particular interest are the existing DICKA protocols [14, 15]. Indeed, they currently lack a tight bound on the relevant entropy, which severely penalizes their performance. Based on the results of [13], one could aim to develop a theoretical framework which enables the derivation of tight entropy bounds for the existing and for future DICKA protocols. This would optimize the security analyses of DICKA protocols and boost their potential application in the upcoming quantum networks.

In the context of DICKA, another potential research line is the characterization of the essential requirements for a successful protocol implementation. To be more specific, in Chap. 7 we conjectured that genuine multipartite entanglement (GME) shared by all the participants is necessary if the inequality being tested is the Mermin-

Ardehali-Belinskii-Klyshko (MABK) inequality [16–18]. It is unclear, however, if the task of DICKA necessarily needs to rely on GME states. Moreover, we raised doubts about the employability of full-correlator Bell inequalities for DICKA protocols. Such open questions could trigger the search for definitive answers which we believe would shed light on novel fundamental aspects of multipartite quantum correlations. In doing so, one could obtain prescriptions that a Bell inequality should fulfil in order to be used in a DICKA protocol, and suggest new DICKA protocols based on Bell inequalities satisfying these conditions.

Experiments on DI cryptographic protocols started to appear only recently [19–23]. Indeed, recent developments in photonic detector efficiencies [24, 25] and parametric down-conversion sources [26, 27] have opened the possibility for an all-photonic implementation of DI cryptographic schemes, paving the way for the application of this technology to future quantum networks.

In conclusion, we hope that this book may support the on-going process transforming QKD and CKA protocols into concrete cryptographic solutions, and at the same time has stimulated further fundamental research on the flourishing field of quantum cryptography.

References

1. Diamanti, E., Lo, H.-K., Qi, B., & Yuan, Z. (2016). Practical challenges in quantum key distribution. *npj Quantum Information, 2*(1), 16025.
2. Murta, G., Grasselli, F., Kampermann, H., & Bruß, D. (2020). Quantum conference key agreement: A review. arXiv:quant-ph/2003.10186.
3. Kimble, H. J. (2008). The quantum internet. *Nature, 453*(7198), 1023–1030.
4. Wehner, S., Elkouss, D., & Hanson, R. (2018). Quantum internet: A vision for the road ahead. *Science, 362*(6412).
5. Broadbent, A., Fitzsimons, J., & Kashefi, E. (2009). Universal blind quantum computation. In *2009 50th Annual IEEE Symposium on Foundations of Computer Science*, pp. 517–526.
6. Fitzsimons, J. F. (2017). Private quantum computation: an introduction to blind quantum computing and related protocols. *npj Quantum Information, 3*(1), 23.
7. Kómár, P., Kessler, E. M., Bishof, M., Jiang, L., Sørensen, A. S., Ye, J., et al. (2014). A quantum network of clocks. *Nature Physics, 10*(8), 582–587.
8. Vaccaro, J. A., Spring, J., & Chefles, A. (2007). Quantum protocols for anonymous voting and surveying. *Physical Review A, 75*, 012333.
9. Wang, Q., Yu, C., Gao, F., Qi, H., & Wen, Q. (2016). Self-tallying quantum anonymous voting. *Physical Review A, 94*, 022333.
10. Grasselli, F., Kampermann, H., & Bruß, D. (2018). Finite-key effects in multipartite quantum key distribution protocols. *New Journal of Physics, 20*(11), 113014.
11. Grasselli, F., Kampermann, H., & Bruß, D. (2019). Conference key agreement with single-photon interference. *New Journal of Physics, 21*(12), 123002.
12. Proietti, M., Ho, J., Grasselli, F., Barrow, P., Malik, M., & Fedrizzi, A. (2020). Experimental quantum conference key agreement. arXiv:quant-ph/2002.01491.
13. Grasselli, F., Murta, G., Kampermann, H., & Bruß, D. (2020). Analytical entropic bounds for multiparty device-independent cryptography. arXiv:quant-ph/2004.14263.
14. Ribeiro, J., Murta, G., & Wehner, S. (2019). Reply to "comment on 'fully device-independent conference key agreement' ". *Physical Review A, 100*, 026302.

15. Holz, T., Kampermann, H., & Bruß, D. (2019). A genuine multipartite bell inequality for device-independent conference key agreement. arXiv:quant-ph/1910.11360.
16. Mermin, N. D. (1990). Extreme quantum entanglement in a superposition of macroscopically distinct states. *Physical Review Letters, 65*, 1838–1840.
17. Ardehali, M. (1992). Bell inequalities with a magnitude of violation that grows exponentially with the number of particles. *Physical Review A, 46*, 5375–5378.
18. Belinskiĭ, A. V., & Klyshko, D. N. (1993). Interference of light and Bell's theorem. *Physical Review A, 36*, 653–693.
19. Liu, Y., Zhao, Q., Li, M.-H., Guan, J.-Y., Zhang, Y., Bai, B., et al. (2018). Device-independent quantum random-number generation. *Nature, 562*(7728), 548–551.
20. Agresti, I., Poderini, D., Guerini, L., Mancusi, M., Carvacho, G., Aolita, L., et al. (2020). Experimental device-independent certified randomness generation with an instrumental causal structure. *Communications Physics, 3*(1), 110.
21. Li, M.-H., Zhang, X., Liu, W.-Z., Zhao, S.-R., Bai, B., Liu, Y., Zhao, Q., Peng, Y., Zhang, J., Zhang, Y., Munro, W. J., Ma, X., Zhang, Q., Fan, J., & Pan, J.-W. (2019). Experimental realization of device-independent quantum randomness expansion. arXiv:quant-ph/1902.07529.
22. Liu, W.-Z., Li, M.-H., Ragy, S., Zhao, S.-R., Bai, B., Liu, Y., Brown, P. J., Zhang, J., Colbeck, R., Fan, J., Zhang, Q., & Pan, J.-W. (2019). Device-independent randomness expansion against quantum side information. arXiv:quant-ph/1912.11159.
23. Shalm, L. K., Zhang, Y., Bienfang, J. C., Schlager, C., Stevens, M. J., Mazurek, M. D., Abellán, C., Amaya, W., Mitchell, M. W., Alhejji, M. A., Fu, H., Ornstein, J., Mirin, R. P., Nam, S. W., & Knill, E. (2019). Device-independent randomness expansion with entangled photons. arXiv:quant-ph/1912.11158.
24. Li, H., Yang, X., You, L., Wang, H., Hu, P., Zhang, W., et al. (2018). Improving detection efficiency of superconducting nanowire single-photon detector using multilayer antireflection coating. *AIP Advances, 8*(11), 115022.
25. Zhang, W., Jia, Q., You, L., Ou, X., Huang, H., Zhang, L., et al. (2019). Saturating intrinsic detection efficiency of superconducting nanowire single-photon detectors via defect engineering. *Physical Review Applied, 12*, 044040.
26. Fedrizzi, A., Herbst, T., Poppe, A., Jennewein, T., & Zeilinger, A. (2007). A wavelength-tunable fiber-coupled source of narrowband entangled photons. *Optics Express, 15*(23), 15377–15386.
27. Graffitti, F., Barrow, P., Proietti, M., Kundys, D., & Fedrizzi, A. (2018). Independent high-purity photons created in domain-engineered crystals. *Optica, 5*(5), 514–517.

Lightning Source UK Ltd.
Milton Keynes UK
UKHW021302091221
395369UK00001B/45